The Fifth Postulate

The Fifth Postulate

How Unraveling a Two-Thousand-Year-Old Mystery Unraveled the Universe

Jason Socrates Bardi

WILEY

John Wiley & Sons, Inc.

Published by John Wiley & Sons, Inc., Hoboken, New Jersey
Published simultaneously in Canada

For general information about our other products and services, please contact our Customer Care Department within the United States at (800) 762–2974, outside the United States at (317) 572–3993 or fax (317) 572–4002.

Wiley also publishes its books in a variety of electronic formats. Some content that appears in print may not be available in electronic books. For more information about Wiley products, visit our web site at www.wiley.com.

Library of Congress Cataloging-in-Publication Data:

Bardi, Jason Socrates.
    The fifth postulate: how unraveling a two-thousand-year-old mystery unraveled the universe/Jason Socrates Bardi.
        p. cm.
    Includes bibliographical references and index.
    ISBN 978-0-470-14909-6 (cloth)
    1. Parallels (Geometry)   2. Geometry, Plane.   3. Pythagorean theorem.   I. Title.
II. Title: 5th postulate.
    QA481.B237 2008
    516.2–dc22

                                                                            2008012187

Printed in the United States of America

10 9 8 7 6 5 4 3 2 1

For Isaac B.

Contents

Acknowledgments

"How do you do it?" a casual acquaintance asked me the other day. I had been telling her about my book—this book—and how I was finishing it while at the same time working full-time at the American Institute of Physics. This was my second book, and the job was my first job in a management position. Both have been tough but rewarding, coming on top of the fact that my wife and I are raising two kids, both under the age of four (Isaac is not yet six months). Again, how do I do it?

The truth of the matter is, I could never do any of these things by myself. There are many people without whose help I never would have done this. Now it is time to thank them.

First and foremost, I would like to thank my wife, Jennifer, for her continuous support, both the obvious help in the form of occasional typing and proofing support and also for her much more subtle and constant assistance. Her deep commitment to me as a writer and as a person is inspiring. Her willingness to adapt her lifestyle to being the wife of an author is humbling—she being a full-time magazine editor as well. There have been many weekends and evenings in the past two years when she took charge of our children and our lives in order to afford me more time to write, and I could not have finished this book without her.

Next, I have to thank my agent, Giles Anderson, with whom I began this book just over two years ago. He has been a tireless advocate throughout the project and a genuine friend. Thanks, too, to Jenny Meyer and her agency for their efforts in conjunction with Giles for the foreign sales of this book.

Special thanks go to my editor, Eric Nelson, and everyone else in the trade division of John Wiley & Sons. They have made this project manageable and rewarding from start to finish. I would like to thank editorial assistant Ellen Wright and all the folks who worked on this book. Thanks especially to Lisa Burstiner, who was a fantastic production editor and gave the book a careful line-by-line read. Also thanks to Roland Ottewell, who was hired by Wiley to copy edit the final manuscript in spring 2008, and he did an excellent job. Thanks also to Hope Breeman, who gave the book its final proofing.

I would like to thank the following friends and family who read the finished manuscript: Lucy DiChiara, John Comeau, Karen Oslund, John F. Bardi, and Phillip Schewe.

Also special thanks to Kevin Fung for designing my Web site and flyers. Thanks, too, to Richard Lerner and all my old friends at the Scripps Research Institute for all the useful discussions, especially Tamas Bartfai, Keith McKeown, and Mika Ono.

Newer friends I have to thank include my coworkers at the National Institute of Allergy and Infectious Diseases, who discussed the work with me in the early stages: Kathy Stover, Courtney Billet, Laurie Doepel, Anne Oplinger, Hillery Harvey, Greg Folkers, Marg Moore, and Sharare Jones. I had some really nice discussions with David Morens on cholera epidemics and Carl Gauss's possible lead poisoning. Also thanks to the members of our now-defunct writing group: Dustin Hays, Linda Joy, and Nancy Touchette.

Even newer friends I have to thank are my colleagues at the American Institute of Physics: Phil, Jim Dawson, Martha Heil, Emile Lordich, Chris Gorski, Dick Jones, Sam Ofori, Sonja Johnson, Tatiana Bonilla, and of course Alicia Torres. Also thanks to Jim Stith and Fred Dylla, who discussed this project with me on a few occasions.

Other friends I have to thank are Teddy and Anna Michele Chao, Nick and Sarah Goffeney, and Johan Hammerstrom and Paula Sanjines and their respective families. Also Albert DiChiara and everyone at the University of Hartford.

Finally, let me say that the moments that define one as a writer are not always the obvious ones—sending off the proposal, signing a contract, sending in the first draft, signing the first copies, depositing a royalty check, and so forth. Certainly these are big moments, but they are like the snow on top of a cold mountain.

Looking at the mountain from a distance, you always see the snow that covers it. Scrape the snow away, however, and you will discover something rocky and complex—like all the frustrations, tiny successes, and endless other moments that make up the long process of writing a book. Ten years from now I will not remember how last night I stayed up late finishing the acknowledgments, but today at least I can reflect back on it and see how this one rock fits into place.

Readers are invited to send comments and other feedback to the author at jasonsocratesbardi@gmail.com.

Prologue

For a brief moment in 1993, mathematics stood where it almost never stands—in the spotlight. A mathematician at Princeton University named Andrew Wiles had just solved a problem known as Fermat's last theorem. This problem was simple to pose but exceedingly difficult to prove, and in 350 years few had ever come close. From the time Pierre de Fermat, a professional lawyer and amateur mathematician, came up with the problem in seventeenth-century France until Wiles found his solution, the theorem had stymied every attempt to solve it—except that of Fermat. "I have discovered a truly marvelous proof, which this margin is too narrow to contain," Fermat wrote in 1637. He claimed, though some have doubted, that he had found a proof almost short enough to fit on one page. Wiles's proof was perhaps a lot more marvelous—pages and pages of equations so complicated that perhaps only a handful of people in the entire world would be expert enough to understand the work.

Still, mathematics stood strong as the news of Wiles's proof exploded over the television and radio. Science writers everywhere hastily sharpened their mathematical pencils and scrambled to describe the discovery in newspapers, magazines, and books. "At Last, Shout of 'Eureka!' in Age-Old Math Mystery," declared the front page of the *New York Times*.

Proving Fermat's last theorem was hailed as the solution to the biggest math problem of the century, perhaps even the millennium. But was it?

A case can be made that the honor of the biggest mathematical solution of the millennium should go to a problem solved more than a century before Wiles was born—one that dates almost all the way back to the birth of mathematics and had dogged mathematicians long before Fermat penned his last theorem. The problem was how to prove Euclid's fifth postulate, his basic statement of parallel lines that states that two lines that are not parallel will cross if they are in the same plane. While it seemed perfectly logical from the moment Euclid scribed the fifth postulate at the beginning of his famous *Elements* in 300 BC, no one from before the birth of Jesus to after the fall of the French monarchy had been able to prove it. Many mathematicians had tried, but every effort ended in frustrating failure.

This is the story of how this mystery was born, with roots reaching back to the earliest Greek mathematicians, Thales and Pythagoras, who converted the mathematics they learned from the Egyptians and Babylonians from a practical tool of urban planning and commerce into an almost mystical and sublime art. It is the story of the enigmatic Euclid and how his unproven postulate persisted for more than two thousand years. It is the story of the generations upon generations of successive Greek, Roman, Arabic, and finally European mathematicians who followed Euclid and struggled valiantly to confirm his postulate. Finally, in the early 1800s, this unsolved problem fell separately to three mathematicians—Carl Friedrich Gauss in Göttingen, Nikolai Lobachevsky in the Russian city of Kazan, and János Bolyai in Vienna.

Working independently, the three men discovered a strange new world called non-Euclidean geometry that would breathe new life into geometry and mathematics. This book recounts the little-known tale of the strangely parallel triumphant and tragic lives of Bolyai, Gauss, and Lobachevsky and how their invention, non-Euclidean geometry, was fostered by the mathematicians who followed in their wake. It is the story of triumph in failure, failure in triumph, and how one of the greatest mathematical problems of the ages finally was solved.

A Mathematician's Waterloo

All the measurements in the world are not worth one theorem by which the science of eternal truths is genuinely advanced.

—Carl Friedrich Gauss

Napoleon Bonaparte was basking in the height of his glory in 1800, and so was another towering figure of the day—the great Italian-French mathematician Joseph-Louis Lagrange. Whether by military or mathematical might, France dominated Europe, and Napoleon and Lagrange were proof of it. In 1800, both were poised to further their mastery. Napoleon seemed set to knock over the rest of the continent, and Lagrange was ready to conquer the entire mathematical world.

In those days, France was flush with great mathematicians. The French Revolution had arrived squarely in the middle of some of the greatest mathematical progress in history. Prior to the revolution, Paris was the center of the mathematics world, and afterward there was an even greater exchange of mathematical ideas. The French capital attracted and educated some of the greatest minds of the day, including Pierre-Simon Laplace, Adrien-Marie Legendre, Siméon-Denis Poisson, Joseph Fourier, Augustin Cauchy, Lazare Carnot, and the young Sophie Germain. Lagrange was the elder statesman among them and the greatest of all.

Lagrange's reputation was hard-won and well deserved. Decades before, as a self-taught teenager, he had worked out a solution to a problem in calculus that had dogged thinkers for half a century despite attempts by some of the greatest minds of his day to solve it.

This solution launched Lagrange's fame, and he never looked back. Elected a member of the Berlin Academy of Sciences, he soon started solving some of the most profound scientific questions of his day.

Lagrange's prowess won him recognition in France almost instantly. He took several prizes offered by the French Academy in the 1760s for his work on the orbits of Earth's moon and the moons of Jupiter. One of the most famous of these writings was his deduction of the so-called problem of libration: why the same side of the moon always faces Earth. Lagrange showed that this was due to the mutual gravitational attractions of Earth, the moon, and the sun and that it could be deduced from Newton's law of gravitation.

By the age of twenty-five, Lagrange had been proclaimed the greatest mathematician alive. That was then. In 1800, he was still great and venerable, but he was an old man, looking back through the window of the French Revolution onto his glorious youth.

Time was ripe for another revolution in mathematics in 1800, and revolution was one thing Lagrange knew quite well. He had been a firsthand witness to the brutality of the French Revolution. Some of his closest acquaintances were put to the guillotine. As far as revolutions went, the French Revolution was premised on a logical, even mathematical, approach to government. Not all of its numbers were pretty, however. The turmoil began after a weak harvest in 1788 brought widespread food shortages in 1789. The panic over food created an untenable political situation that came to a head during the summer, when crowds stormed the Bastille, which, in addition to being a prison, had become a repository for gunpowder. This led to the collapse of the monarchy and the rise of a constitutional assembly, which within a few years declared a republic, passed a slew of new laws, and tried King Louis XVI for treason, chopping off his head in 1793.

In the year or so that followed, known as the Reign of Terror, more numbers became apparent. Some 2,639 people were decapitated in Paris. Thousands more lost their lives as mass executions played out across France. The horrors of the guillotine are well known, but this was not the only method of slaughter. In the city of Nantes, the victims of the Terror were killed by mass drowning.

During the Reign of Terror, Lagrange was in a precarious situation. He was a foreigner without any real home. He is said to have been an Italian by birth, a German by adoption, and a Frenchman by choice. His roots were in France—his father had been a French cavalry captain who entered the service of the Italian king of Sardinia, settled in Turin, and married into a wealthy family. Lagrange's father was much more adept at enjoying the proceeds of his rich lifestyle than at maintaining his wealth, however. A lousy money manager, he wound up losing both his own fortune and his wife's before his son would see any of it.

Lagrange left home early to seek his fortune, and he found it in the complex, abstract, and imaginatively free world of numbers and math. He rose to become a famous mathematician—the most famous. Frederick the Great appointed young Lagrange to be the director of the Berlin Academy in 1766, and he spent many of his best years there. After Frederick died in 1786, Lagrange had to leave Berlin because of anti-foreign sentiment, so he accepted an invitation by Louis XVI to come to Paris and join the French Academy of Sciences. He took up residence in the Louvre and became close friends with Marie Antoinette and the chemist Antoine Lavoisier.

Lagrange was a favorite of Marie Antoinette's. On the surface, his was perhaps as enviable a position as a mathematician of his day could hope for. But he fell into depression and decadence and became convinced that mathematics, too, was shrinking into decadence. Then the French Revolution arrived. Lagrange could have left at the outset, but where would he go? Not back to Berlin—and not back to Italy, a country he had left as a young man and to which he was no longer connected. For better or worse, France was now his home.

Lagrange lived to regret his decision and almost lost his life when the Reign of Terror started. When he was facing the guillotine, he was asked what he would do to make himself useful in the new revolutionary world. He insisted that it would be more worthwhile to keep him alive. To avoid being put to death, he replied, "I will teach arithmetic."

Napoleon was already a rising star in France when he seized power and began setting up a new state of his own making with himself at the top. He had a keen interest in the educational system, which

meant that he soon began to take a keen interest in Lagrange. He
selected Lagrange to play a leading role in perfecting the metric system of weights and measures. By the dawning of the new century,
Napoleon had come to refer to Lagrange as his "high pyramid of
the mathematical sciences."

The year 1800 was a new day, both for France and Lagrange.
The French had defeated the Dutch, crushed the Prussians, and
annexed Belgium. There would soon be a pause followed by an
even more explosive period of warfare. Mathematics was a changing
field as well. It was quickly becoming more international, and
nobody embodied this more than Lagrange. Napoleon made him
a senator, a count of the empire, and a "Grand officer of the legion
of Honor." He rose higher and higher. Napoleon often consulted
with Lagrange between campaigns—not for military advice, but for
his perspective on matters of state as they related to philosophy
and mathematics.

Lagrange began to lead France's two great academies, the École
Polytechnique and the École Normale Supérieure, and was a professor of mathematics at both institutions. For the next century
all the great French mathematicians either trained there or taught
there or did both. This was where Lagrange reigned supreme. In
1800, nearly sixty-five years old, he was the premier mathematician
in France, not through his acquaintance with Napoleon but by his
dominance of the field through the previous two generations. He
was positioned to profoundly influence legions of young mathematicians through his teaching and his original and groundbreaking
work developing methods for dealing with rigid bodies, moving
objects, fluids, and planetary systems. Lagrange's greatest discoveries were behind him in 1800, but he was perfectly positioned
to reclaim his past glory. He was ready to make history again—and
this time more dramatically than ever.

Such was the mood one day in 1800, when Lagrange stood up
in front of an august body of his French peers, cleared his throat,
and prepared to read what he must have thought would be one
of his greatest breakthroughs. He was about to prove Euclid's fifth
postulate—the mystery of mysteries. The oldest conundrum in
mathematics, it dealt with the nature of parallel lines. Proving it had
dogged mathematicians for thousands of years. Lagrange's own life
was a microcosm of this history. One of his earliest encounters with
mathematics was the work of Euclid, the ancient writer who had first

proposed the fifth postulate thousands of years before, in 300 BC, in his treatise *Elements*. Lagrange stumbled upon this problem as a boy. He was aware of it for his entire career.

Euclid could not solve the fifth postulate, nor could the ancient Greek and Roman thinkers who followed him, nor the Arabian scholars who translated Euclid's work into their own language, nor the Renaissance intellectuals who translated the work into Latin and the European languages and studied it at their universities, nor the visionaries of the scientific revolution who developed mathematics as never before, nor finally any one of the many mathematicians who surrounded Lagrange. Some of the best minds of the previous twenty centuries took a stab at proving it, but they all missed the mark.

The audience surely crackled with anticipation as Lagrange took the podium to read his proof. The beauty of his work was well known. A dozen years before, he had published a book called the *Méchanique analytique*, which became the foundation of all later work on the science of mechanics. It was so beautifully written that Alexander Hamilton called it a scientific poem. A century later, it was still considered one of the ten most important mathematical books of all time.

Who in the audience could question that proving the fifth postulate was to have been Lagrange's greatest discovery yet? No mathematician in the previous two thousand years had been able to do it—a period that included Archimedes, Isaac Newton, and everyone in between. Some of the greatest thinkers in the history of mathematics had tried and failed. There was nearly a continuous chain of failed proofs stretching all the way back to Euclid, and it probably went back even further. Nobody knew for sure when mathematicians had started trying to prove the fifth postulate, and nobody in the audience that day could possibly account for how many had tried through the years. The only thing that was clear on the day that Lagrange stood ready to prove the fifth postulate in 1800 was that all before him had failed.

A postulate is a statement without justification—not in the sense that it is absurd but in the sense that it is not or cannot be proven. The fifth postulate basically says that two lines in a plane that are not parallel will eventually cross. Nothing seemed more obvious. Nevertheless, nobody had proved this to be true in all cases.

In Lagrange's day, the fifth postulate was called the scandal of elementary geometry, and working on it was the height of fashion for the cream of mathematicians in Europe. There had been a flood of papers on it in previous years; all aimed to prove it and all failed. There was nothing murkier in math.

How excited must Lagrange have been as he stood up to speak? Perhaps he saw this as a defining moment in history, like Napoleon marching into Germany, Poland, and Spain, his heart thumping like the constant drumbeat of marching armies. But to Lagrange's great embarrassment, he suffered a mathematical Waterloo instead. He made a simple mistake in his proof, and many in the audience recognized it immediately. Then he saw it himself, and abruptly ended his presentation, declaring, "Il faut que j'y songe encore" ("I shall have to think it over again"). He put his manuscript in his pocket and left. The meeting went on with further business.

History vindicates Lagrange somewhat. He was the last of a string of frustrated mathematicians who for thousands of years had been trying to prove the fifth postulate directly. They were like mountain climbers trying to scale the highest peak. But they would only ever get so high before they came to a chasm. Proving the fifth postulate was like trying to find a way across this chasm.

It *had* to be true, and it *had* to be proven. The fifth postulate was not some obscure mathematical concept. Euclidean geometry was no mere mathematical treatment of abstract ideas. It cut to the very nature of space itself. Euclid's *Elements* was irreproachable. It was seen as the guidebook to truth in elementary geometry, and geometry was a treatment of reality—the space around us. To read Euclid was to know geometry, and to know geometry was to know reality. As the book was studied and translated through the years, numerous notes and remarks were compiled by various scholars who found ways to explain certain parts of the book. More than a thousand of these accumulated, but none of them could ever successfully address the one remaining unsolved problem.

Who could possibly question the reality of physical space? Certainly not Lagrange or anyone before him. The problem was not with his ideas. The problem was with geometry itself. Lagrange failed for the same reason that all mathematicians for centuries had failed: the fifth postulate could not be proven. But where Lagrange failed, three mathematicians would soon succeed.

In 1800, the world knew nothing of Carl Friedrich Gauss, Nikolai Lobachevsky, and János Bolyai, and the three knew nothing of one another. They would never meet, but they all shared at least one obsession throughout their lives: solving the fifth postulate. If ever there was a mystery calling out for a fresh approach, this was it. Working independently, Gauss, Lobachevsky, and Bolyai would each climb the same mountain and stare down the same chasm. Any effort to prove the fifth postulate was a bottomless pit, and even if they poured a lifetime of effort into it, they never would have reached its floor. Realizing this, many mathematicians had given up trying to find proof. But Gauss, Bolyai, and Lobachevsky all took a new approach. They would solve the mystery of the fifth postulate by asking a completely different question: what if the postulate was not true at all? They would first begin determining what space might be like in their alternative geometry. This would give them insight into the nature of three-dimensional space that few could imagine—certainly not Lagrange in 1800. The reason Lagrange and all the other mathematicians in history could not find this solution was that it required a leap of faith that none of them was ready to make. The solution to the fifth postulate lay in rejecting it entirely and creating a whole new world of geometry.

This new world was given many names in the nineteenth century—astral geometry, imaginary geometry, absolute geometry, hyperbolic geometry—and finally became known as non-Euclidean geometry. It is one of the great achievements of the human mind. It was as if for two thousand years mathematics was an orchestra composed entirely of drums. Mathematicians were like composers seeking to write the arrangement, but they were limited by the instrumentation. Then these three mathematicians came along and examined what music would be like if they were not constrained to the drum. They then invented the piano!

Instead of solving the oldest problem in Euclidean geometry, these three mathematicians invented non-Euclidean geometry. In doing so, they opened up the mathematical orchestra to millions of new arrangements, new problems, and new ways of looking at space. Non-Euclidean geometry was not a correction but a whole new geometry that introduced a strange space in which straight lines are curved and geometric objects become more distorted the larger they are. The oddest thing of all was that the strange new world turned out to be correct.

As Lagrange retreated, embarrassed, from the podium, how could he have known that the real answer to the mystery he had just failed to prove was already in the head of the young Gauss, who was just out of university? How could he have guessed that years later it would be discovered by Lobachevsky, who was then only a boy growing up in a remote part of Russia? How could he have even imagined that Bolyai, who was not even born in 1800, would also come up with non-Euclidean geometry on his own?

After these men formulated their theories, mathematics would never be the same.

Carl Friedrich Gauss coined the term *non-Euclidean,* but he never published anything on the subject in his lifetime. He was without a doubt the greatest mathematician of his day, perhaps one of the three greatest mathematicians who ever lived, so it is strange that one of his greatest discoveries went unmentioned until after his death.

Gauss was also an odd creature, so obsessed with numbers and numerical relationships that he could recall numerical solutions to complicated arithmetic problems he had solved years before. He used to keep many little notebooks filled with numbers, including days of historical note, biblical references, observations, and calculations. He even had one with the dates when his children's teeth came in. For amusement he would record the ages of all of his children and many of his friends in days. A few nights before he died, as if responding to a premonition of his own imminent demise, he did the same with his own age.

Gauss's obsession with numbers was not part of some grand delusional scheme trying to find meaning in a swarm of numerical nonsense. It was merely for enjoyment. Though to an unappreciative observer Gauss may have seemed like a madman, he really wasn't crazy at all. He was just firmly entrenched in numbers and math.

He was a mathematical super-genius—to call him a mathematical genius is to sell him short—possessed of an all-encompassing and deeply penetrating vision that mere genius never approached. Gauss claimed that he learned to do arithmetic before he could walk. He is said to have been able to do cube roots in his head by the age of eight. And he once told one of his students, "You have no idea how much poetry is contained in the calculation of a logarithm table." In fact, he enjoyed using mathematical tables that were cheap and inaccurate so that he could correct them as he went.

Of course, not everything he did was so trivial. He also tackled some of the most important questions of his time and stretched his work across every field of mathematics, from number theory to geometry to probability to analysis. And he made significant contributions to related fields, such as astronomy and physics. He was so successful that many of his contemporaries considered him the greatest mathematician in the world even when he was a young man.

Among other innovations Gauss brought into the world, he introduced the "normal" distribution—the Gaussian, or so-called bell curve that represents how observed data are distributed. The bell curve was a way of representing statistically distributed data, and this distribution represented the important concept of probability. Gauss was the first person to use the letter i as a notation to indicate imaginary numbers, a standard still in use today, and he came up with the term *complex numbers* to describe those numbers that have both a real and an imaginary component.

Most of all, Gauss is remembered for his stunningly complete and sometimes perfect works of mathematical art. Around the same time that Lagrange was embarrassed in his attempt to prove the fifth postulate, he read Gauss's first book. The old mathematician was profoundly impressed with his much younger contemporary. "Your *Disquisitiones* have with one stroke elevated you to the rank of the foremost mathematicians," Lagrange gushed in a letter to Gauss, also referring to one of the discoveries in the book as "one of the most beautiful which has been made in a long time."

Gauss spent his entire life chasing mathematical masterpieces. He considered everything else to be just scaffolding—and often unworthy of publication. He regarded himself in such poetic terms as well. He had Shakespeare's words from *King Lear* inscribed underneath a famous portrait of him by the artist Christian Albrecht Jensen: "Thou, Nature, art my goddess; to thy laws my services are bound."

Gauss's devotion to perfection may have stalled mathematics in his lifetime, however. He never published several of his discoveries, and more than one great mathematician of the nineteenth century discovered that some of their greatest work had been previously discovered by Gauss but remained unpublished in his notes. This was also the case with non-Euclidean geometry. According to his own recollections later in life, Gauss was a teenager when he started to figure it out. Throughout his lifetime, though, he rarely revealed even a hint about his thoughts.

Politically, this was a time of change in Europe. Revolution and reaction and their consequences were touching the continent. This would have been an interesting time for Gauss to shake up the mathematical world by exploring and revealing his strange new world of geometry. All he had to do was to publish something—anything. Instead, he said nothing.

When Gauss was a boy, he nearly drowned when he fell in a canal near his house. He was rescued, and nobody could have guessed how close the world came to suffering the cruel fate of losing one of its greatest minds.

Nobody thought much of the boy when he was a child. Born Johann Friedrich Carl Gauss, he went by Carl Friedrich Gauss his whole life. He was the only son of his impoverished parents, Gebhard Dietrich Gauss and Dorothea Benze, and was his father's second son. He had an older half-brother from his father's previous marriage. Gebhard's first wife died two years before Gauss was born.

Gauss's mother, Dorothea, was loving and devoted, but Gauss enjoyed few other advantages in childhood. To say his family was of simple means would be to put it mildly. He was born into severe poverty, at a time when the harsh misery of such an existence often bore down on those in his station. His grandfather was a stonemason and died at the age of thirty, having ruined his body by breathing in dust from the sandstone he worked with. His father lived slightly longer than that, and his mother lived to a ripe old age, but their lives were not much easier.

His paternal grandparents were subsistence farmers who sought to improve their fortunes by moving to the German city of Brunswick as "half citizens"—part of an overall influx of immigrants to the city at the end of the eighteenth century. Gauss's father worked a series of dead-end jobs. He was a street butcher for a time and a gardener for another. He was a canal tender, then became a part-time bricklayer, his occupation when Gauss was a child.

When Gauss was born, his family was so poor that nobody bothered to make an official record of his birth. He was a commoner and had his childhood documented as befitting one of his station—which is to say nearly not at all. He didn't know when his birthday was until he was an adult—and then only because he figured it out himself. Even his mother could not remember his exact birthday, but she did recall it was a Wednesday and that it was a certain number of days removed from Easter. So when Gauss was a young man, he sat down one day and figured out a simple calculation to determine on which weekend Easter would fall in any given year between 1700 and 1899. Then he worked backward and determined his birthday.

The technique involves dividing the year by a series of numbers, doing another set of calculations with the answers, and then adding the remainders together with the number 22 to give the date of Easter. That Gauss had the patience and discipline to come up with a simple general calculation to solve this problem is incredible. Most people then, as now, would probably hasten to a library or some used bookstore and simply look up in an old record to see on which date Easter fell in 1777, and then figure out their date of birth from there. Gauss used his calculation to determine that he was born on April 30, 1777.

But this was a characteristic of Gauss. This simple boy of humble means would go on to accomplish great things. He would use mathematics to erase the brutality of his impoverished background in his escape from the misery his parents knew. He seems to have completely succeeded in doing so. In fact, as an older man, he was given to recalling fondly the simpler times of his childhood.

Even as a boy, Gauss's mathematical genius appeared. One Saturday when he was three, for instance, his father was dividing up the pay among the other bricklayers. With his son listening intently, he calculated the divided amounts. Gauss supposedly interjected and told his father that the calculation was incorrect. His father was often strict, harsh, and brutal. But he must have softened when, to the astonishment of all, the boy was right.

Gauss was the type of boy who must have seemed a little weird, even to the people who loved him. He was driven to learn and would spend hours doing nothing but reading books. His father, who was thrifty, made him and his brother go to bed as soon as it got dark so as to conserve candles and lamp oil and to save money by not having to heat the house. This did not prevent Gauss from studying his mathematics, though. The boy would sit for hours in the dark reading by the light of animal grease rendered from some carcass, which he had burning in a hollowed-out potato or turnip that he filled with the fat and lit with a homemade wick. He would read for hours by this putrid light until he fell asleep. These flickering flames were enough to illuminate page after page of mathematical text that the boy would pound into his skull every night.

This was a remarkable difference between Gauss and his parents. They were simple, semiliterate folk—if that. Their meager education gave them the barest tools for dealing with numbers, while Gauss received one of the finest educations that could be had in his day and grew to have an incredible facility for manipulating both words and numbers.

In some ways, this transformation from one generation to the next was amazing. Gauss was able to advance himself completely and escape the world of his parents entirely. His brother was a weaver, his father a bricklayer. Gauss's rise from son of an illiterate laborer to one of the top academics of his time was an unusual path for a boy of his station, almost miraculous. Certainly his parents had no expectation that his schooling would amount to anything spectacular. Perhaps they might have hoped for him to become a schoolteacher, maybe even a merchant. They never would have expected him to go to college and university for the next decade and a half and become the greatest mathematician of his day, because such an opportunity didn't exist for people like his parents.

Gauss's own brother thought of him as good for nothing because he always had his head buried in a book. And the boy might very well have been pulled early and permanently away from mathematics to live out his life as a meager peasant, breaking his back at his humble trade for pennies and dying completely forgotten.

However narrow his parent's view of the doors his education could open, they sought to encourage the boy because he definitely had a gift for numbers. At the age of seven, Gauss entered the nearby St. Katherine's school in Brunswick. Although only a small percentage of Germans kids went to school in those days, schools were becoming more accessible to the general populace, more than doubling the literacy rate in Germany over the eighteenth century. These gains were not always extended beyond the basic levels, however, and few continued after the first few years of elementary studies. Gauss might have stopped early too, except that he had the good fortune to impress the right people at the right time. He had that extremely rare combination of a unique mind and rare luck—much as he had been lucky when he was by chance seen falling into the canal by his home and plucked from a watery death. Gauss lived, and he drowned himself in numbers instead.

Two hundred students were crammed in a creaky, musty room at St. Katherine's school as their master, Mr. Büttner, stalked the aisles with a whip in his hand dispensing the cruel lessons of his curriculum. Incorrect answers would be met with lashes. Many of Büttner's students probably had whatever fleeting interest they might have had in mathematics and learning in general whipped out of them by those lashes—but not Gauss. He saved himself from the whip by showing flashes of genius, and his experience ignited a lifelong love of mathematics.

The day Mr. Büttner realized young Gauss had a real gift was probably one like any other. He assigned the children a bit of busy work: adding up every integer between 1 and 100. The students got to work, and Mr. Büttner monitored. Most of the students probably solved this problem in the straightforward manner: sequentially adding up all of the numbers in the set by adding each number to the growing sum:

$1 + 2 = 3;$
$3 + 3 = 6;$
$6 + 4 = 10;$
$10 + 5 = 15;$
$15 + 6 = 21;$
and so on.

How likely are you to make a mistake in any one of these operations? Perhaps minimally, if you know how to add your numbers well—unless you factor in the stress of doing all the additions while at the same time listening to Mr. Büttner as he crept up the aisles. Then you have to consider the number of additions you are making. Adding a hundred numbers in the straightforward manner like this requires ninety-nine operations, and long runs of addition are problematic because ninety-nine operations means ninety-nine chances to get the sum wrong. The chances of making at least one mistake in a hundred are understandably much larger. With the number of additions and the pressure of Mr. Büttner's whip and the cacophony of two hundred children furiously slapping chalk to two hundred slates, it is a wonder that any of the children could get the problem done, let alone correctly. All of the students made mistakes—except for one.

Gauss simply wrote a single number on his slate and immediately put it down. While many of the other boys in the class took a much longer time carefully adding up all the numbers, only to get the sum wrong in the end, Gauss had the answer in mere seconds.

While Mr. Büttner was waiting for the rest of the children to finish, he noticed Gauss sitting in his seat not working. What doubt and scorn he must have felt for the boy! Was little Gauss mocking him? He continued to pace the aisles as the other boys slowly finished their work on the problem. But when he checked the answers, he was surprised to see a single number written on Gauss's slate—shocked, moreover, to find that it was the correct answer of 5,050. How on earth could the boy have guessed the correct answer like that?

In fact, Gauss didn't guess at all. He calculated the answer. What accounted for his speed and accuracy was that he approached the problem differently. Although it appears to require ninety-nine additions of one hundred numbers, the problem is actually much simpler. It can be solved with a simple multiplication of two numbers. Gauss saw that you could add $1 + 100$, $2 + 99$, $50 + 51$, and similar pairs of ascending and descending numbers together. All pairs added up to a sum of 101, and there were exactly 50 such pairs. Thus adding together all the numbers between one and 100 was simply a matter of multiplying 50 and 101, a calculation Gauss found simple enough to do in his head.

To Mr. Büttner's credit, he recognized that this simple little son of a bricklayer had a genius that would have to be fostered. He ordered a more advanced arithmetic book for the boy, and his assistant, Johann Christian Martin Bartels, took a special interest in Gauss and began tutoring him. Bartels was eight years older than young Gauss. He grew up in the same area, lived close by, and was very interested in mathematics himself. Bartels was perhaps the last person Gauss met who knew more about mathematics than he—and only because as an untraincd boy, Gauss knew so little at the time, not because Bartels was a particularly knowledgeable scholar. For the rest of Gauss's life, he would have contemporaries—collaborators. But he would always be the master and never again the pupil. Still, Bartels taught a lot to the boy in these sessions.

Working with Gauss on basic mathematics profoundly influenced Bartels as well. After this experience, he began a love affair with the subject that would last the rest of his life. Later in his life, he became a respectable mathematician who had a decent career as a teacher, and he published a number of essays before he died. Bartels profoundly influenced the development of non-Euclidean geometry because he taught two of its inventors how to do mathematics—both Gauss and later the Russian mathematician Lobachevsky, who invented non-Euclidean geometry independently.

Together, Büttner and Bartels also helped free up Gauss's evenings for private study. As a poor child of simple means, he had to do what is now known as piecework most evenings in order to supplement the family's meager income. These were the sort of time-consuming, mindless weaving or assembly tasks that could be farmed out to a boy. His father had a spinning wheel set up in the house, and every evening Carl had to spin a certain amount of flax yarn, which was used to make twine and fishing nets.

But Büttner and Bartels convinced Gauss's father to release the boy from this busywork and let him study instead. They were so convincing that not only did Gauss's father agree to release his son from further burdens and forgo the modest addition to his income, but as if to emphasize this point, he carried the spinning wheel out back, picked up an ax, and chopped it into firewood.

By far the most profound influence Bartels had on Gauss was to bring him into contact with the people who would really make a difference in the young boy's future—the local nobles who could afford to send the lad to better schools and pay for him to concentrate on mathematics. There was little hope of ever being able to succeed without the intercession of a rich, noble benefactor, and Bartels brought Gauss to the attention of just such a man: Eberhard August Wilhelm von Zimmermann.

Zimmermann was a professor and a close adviser to the local ruler, Duke Karl Wilhelm Ferdinand of Brunswick-Wolfenbüttel. Bartels told Zimmermann about Gauss, and Zimmermann brought the boy to the duke's attention. In those days it was fashionable for noble lords like the duke to foster genius within their realm to adorn their courts with mathematicians and other scholars. It was not unusual for a nobleman such as Duke Ferdinand to sponsor a brilliant peasant such as Gauss.

Duke Ferdinand's initial overall impression of Gauss, from the report he received from Zimmermann, was enhanced by the duke's wife, who happened upon the boy reading a book one day in the palace yard. She walked up to him and asked him about it and was immediately impressed that the thoughtful boy had such an in-depth understanding of the subject. She went back to the palace and urged the duke to summon the boy back.

The duke sent one of his men to fetch Gauss, but a misunder-standing caused the man to mistakenly invite Gauss's older brother instead. George Gauss, realizing this was a mistake, sent the messen-ger along to see his brother. He thought little of the invitation and had no desire to appear before the duke. Years later, however, George would be filled with regret over this when his little brother was a world-famous mathematician. "If I had known," he said, "I would be a professor now." Perhaps he would have become one—if he had been as gifted a thinker as his half-brother. We will never know. What is clear, though, is that young Carl Gauss was some-one extraordinary indeed.

In the nineteenth century, there were plenty of parlor-trick mathematicians who became famous for their ability—usually as children—to multiply large numbers in their heads quickly. One boy named Zacharias Dase, who lived at the same time as Gauss, was famous for doing things like computing the square root of a hundred-digit number in less than an hour and multiplying two twenty-digit numbers in six minutes flat. Other mathematical whiz kids would travel as sideshow attractions and perform similar com-plicated arithmetic tricks for the amused crowds. Such prodigies were usually children, because they often lost their ability to do the tricks as they grew older. In the age before the Internet, before com-puters, and before electronic calculating machines of any sort, they were the original wunderkinds of the mathematical world.

Gauss was no mere amusement, though. He certainly could per-form outstanding feats of mathematics in his head, but his talent was much deeper and more creative. Throughout his life, he sought, and often succeeded, in tackling problems that required months of study or more. By the time he was eleven, he was already a great mathematician. He began to construct and use what are known as infinite series to solve problems. He also worked out a general form of the binomial theorem—something that had puzzled the best mathematicians in a previous century until Isaac Newton discovered it. Newton was older and better educated when he did his work. Gauss was still a boy.

Gauss, on the strength of his promising progress, was rewarded by the duke with a lavish education far beyond anything his parents could have afforded. Under the duke's patronage, Gauss enrolled in Brunswick's prestigious Collegium Carolinum in 1792, and then in the University of Göttingen in 1794. Duke Ferdinand paid for his tuition, provided him with books, and supported him with a modest stipend from his personal fortune after Gauss graduated. The duke supported Gauss even when his state treasury was on the brink of bankruptcy and, as one of his biographies reads, enabled Gauss to permanently exchange "the humble pursuits of trade for those of science."

Gauss was forever grateful to his benefactor. "I owe to your kindness, which freed me from other cares and permitted me to devote myself to this work," he wrote to the duke. "For if your grace had not opened up for me the access to the sciences, if your unremitting benefactions had not encouraged my studies up to this day . . . I would never have been able to dedicate myself completely to the mathematical sciences to which I am inclined by nature."

The end of the eighteenth century was an exciting time for mathematics, and at the Collegium Carolinum, Gauss read the works of Euler, Lagrange, Newton, and all the great mathematicians of earlier generations. He also began a serious study of geometry and became acquainted with the fifth postulate.

More than fifty years later Gauss would tell a friend that in 1792, when he was fifteen and studying at the Collegium Carolinum, he first realized the basis for non-Euclidean geometry. The foundation of his idea later in life was that if the fifth postulate were not true, there would still be a consistent geometry. In other words, he thought he could reject the fifth postulate and still consider valid geometrical concepts. But it is hard to say concretely what his earliest concepts were because he never published them. Nor are there handwritten

notes or any other indication of his views. Still, it was incredible that
he was doing this at a time when the rest of the mathematicians in
Europe, like Lagrange, were still trying to prove the postulate.

From the Collegium Carolinum, Gauss matriculated into the
University of Göttingen, a school founded by King George II of
England and built upon the models of Cambridge and Oxford. It
had a nice endowment and an equally generous amount of academic freedom. It quickly became one of the most prestigious institutions in Germany and remained so for many years. When Gauss
went there, the school was enjoying some of its greatest years and
was attracting students from all over Europe. Even scholars from as
far away as the United States came to visit the university. Benjamin
Franklin, for instance, visited Göttingen when he was formulating
plans for the University of Pennsylvania.

Despite his apparent interest and his obvious talent in mathematics, however, Gauss almost didn't pursue the subject when he
arrived at Göttingen. He had a gift for words as well, and he might
have easily followed something else like law that would put him on a
safer path to a more secure income. Humanistic studies were excep-
tionally strong at Göttingen, and it would have been a natural
fit. What turned him around was his first mathematical discovery—
also one of his most curious. He discovered how to construct a
seventeen-sided figure with only a ruler and a compass.

On March 29, 1796, the day Napoleon left Paris for Italy, where he
would lead French troops on his Italian campaign, win the acclaim
of the empire, and earn the moniker "the little corporal," Gauss
awoke in the morning thinking about regular polygons—shapes
like triangles, hexagons, and pentagons where all the sides are the
same length and all the angles formed by the adjoining sides are
equal.

Since ancient times, mathematicians had been able to construct
regular polygons of a certain number of sides. Shapes like the triangle (three sides), square (four), pentagon (five), hexagon (six), and
octagon (eight) were all known from the time of the ancient Greeks
and before. For millennia, mathematicians, even schoolchildren,
had learned to construct such shapes. There seemed to be a limit,
however, to how many sides a polygon constructed this way could
have.

"It has generally been said since then that the field of elementary geometry extends no farther," Gauss wrote in a German journal in 1796. "At least I know of no successful attempt to extend its limits." Extending the limits of geometry is exactly what he did.

In the history of the world, nobody had ever been able to make a septadecagon, a regular seventeen-sided polygon, with only a ruler and a compass. It was not that nobody had thought of constructing such a figure. In fact, mathematicians had attempted to do this since ancient times. Nineteen-year-old Gauss discovered how. As he later wrote, "After intense consideration of the relation of all the roots to one another on arithmetical grounds, I succeeded on the morning of [that] day before I had got out of bed."

Gauss did much more than construct a seventeen-sided figure, however. He also solved the problem generally and figured out a formula for determining larger polygons that could also be constructed with a ruler and a compass. In fact, he worked out that a polygon with a given number of sides could be constructed *if* the number of sides is a prime number equal to something known as the Fermat number $2^{2^n} + 1$. This meant that he could construct polygons with a number of sides equal to 3, 5, 17, 257, 65537, . . .

If a seventeen-sided polygon had been so elusive, it is unimaginable that anyone had thought of constructing a 257-sided polygon, let alone a polygon with 65,537 sides.

The seventeen-sided figure delighted Gauss. He always considered it one of his most important discoveries. As mentioned earlier, he kept a notebook of significant dates of importance in his lifetime, such as when someone spotted a new planet or when he had a breakthrough mathematical insight. The discovery of the seventeen-sided polygon was one of those dates. So highly did Gauss regard this figure that he told his friends that he wanted to decorate his tombstone with it. When he died more than half a century later, it did indeed become the thematic adornment on the back of the monument erected in his honor. (It was actually a seventeen-sided star, because the stonemason thought a seventeen-sided figure would look too much like a circle.)

Perhaps one of the reasons that the septadecagon figure was so gratifying to Gauss was that it turned him permanently toward mathematics. It was, in a sense, this discovery that launched his long and amazingly productive career. He never looked back, and by the time he died he was considered the "prince of mathematics," the most famous mathematician of his day and one of the three greatest of all time.

Gauss may have been the last mathematician to contribute to every field of mathematics that existed in his day, including geometry. He made some of the most significant discoveries of his time in geometry—not the least of which was the solution to that burning question of the fifth postulate, which Lagrange and many others had failed to prove by the end of the eighteenth century. In those days, while Lagrange was preparing to read his failed proof in Paris, Gauss began to acquire an almost unique lack of faith in the fifth postulate. Instead of asking how to prove it, he began to question why it needed to be proven at all.

Almost nobody had asked questions like this before Gauss, because the fifth postulate was a central part of geometry, and geometry was an established subject communicated through the centuries as the wisdom of the ancients. The fifth postulate was part of this ancient and incontrovertible tradition handed down through the ages and studied for thousands of years in the form of Euclid's text on geometry, the *Elements*.

The *Elements* was one of the most successful textbooks of all time; it survived the rise and fall of the Roman Empire, persisted in the Middle East, and was translated into Latin, Arabic, English, and a half dozen other languages, becoming a standard source for centuries—even before the advent of the printing press. The next-longest-running scientific book, Ptolemy's *Almagest,* was only in use from the second century AD to the late Renaissance.

After the printing press arrived in Europe, the *Elements* was the first mathematical book to be pressed and bound, and on average about two editions were printed each year after that. By the time Gauss was born, more than a thousand editions had already been printed, and the book had been translated into all the modern European languages. The *Elements* is perhaps second only to the Bible in this regard, and like the Bible, it had become more than just lines on a page. It was the embodiment of a certain type of knowledge—a guide to analytical thinking and the scientific method through the ages.

Its author, Euclid, was also unassailably famous. He is perhaps the only mathematician ever who captured his entire field in one book. Because of the book's endurance, Euclid is sometimes considered the leading mathematics teacher of all time. He was so famous that some Arabian writers claimed that he was really an Arab, the native son of Tyre.

In fact, Euclid was a native son of Alexandria, Egypt. And geometry was not his unique invention, but something he invented in part and borrowed, stole, and largely cobbled together from other, much earlier sources. Centuries of Greek mathematicians who came before Euclid really deserve much of the credit for the *Elements*. Hundred of years of discoveries by people who have been partly or completely lost in time—men and women who witnessed the flower of Greek civilization and are now forgotten—contributed in some way. All of these discoveries found their way into the *Elements*.

Euclid's *Elements* also owes a lot to people who will be remembered forever—like Plato and Aristotle. Plato was not interested in experimentation and regarded it to be a base art. Mathematics, on the other hand, he appreciated fully. He was influenced by one of his teachers, Theodorus of Cyrene, who taught him mathematics. Plato's philosophy was a mathematical one, and he brought a love of mathematics to learned circles that persists to this day. His academy endured in one form or another for more than nine centuries. Moreover, the idea of the academy as a place of learning would greatly influence the course of learning in general and mathematics in particular.

Aristotle was even more influential. He seems to have been the first to consider organized research and the first to classify knowledge into the different disciplines that existed in his time. He was the first to attempt to formally organize science into a logical approach, and for nothing else, Aristotle would be remembered for his large contribution to the Western culture of improving the systematic presentation of mathematics and other subjects.

Aristotle is said to have derived his theory of science from the very fruitful system of mathematics that existed during his lifetime. He organized his approach to science the way that mathematics was organized—as a system of laws derived by logical approaches to proof based on a minimal number of assumptions. Science, like mathematics, took simple definitions and postulates and built upon them.

Plato and Aristotle were both alive slightly before Euclid, so they never saw his masterwork, but they were likely familiar with the same books that Euclid used to write much of the *Elements*. And Euclid was deeply influenced by them. He took the axiomatic system of Aristotle and furthered it, adapting it and applying it to mathematics in his book.

Euclid's book also owes something to the most famous of the ancient Greek mathematicians—Thales and Pythagoras and their followers. They lived hundreds of years before anyone had ever heard of the fifth postulate, and they imported mathematics from Egypt and Babylon in the seventh century BC. The mystery of the fifth postulate really began with them.

2

The Strange Vegetarian Cult and Mathematics

[Pythagoras] easily beheld everything, as far as ten or twenty ages of the human race.

—Empedocles, quoted by Iamblichus of Chalcis,
a fourth-century Neoplatonic philosopher.

According to ancient legend, Pythagoras was the son of Apollo, half god, half man—a sort of Hercules, only a whole lot smarter. In a sense, he was actually two different people. There was Pythagoras the man, a real person who was born around 580 BC, roamed about the ancient world, and died eighty or ninety years later. Then there was the Pythagoras of legend—Pythagoras the god—whose accomplishments were much greater than any ordinary man's.

To this man are ascribed almost magical feats of mathematics. One time he supposedly happened upon some fishermen hauling a large catch in their nets, and he boasted that he could determine the exact number of fish they had caught. The fishermen, perhaps surprised at the boast, played along and said that if he could guess the number correctly, they would do anything he said. So he guessed. And they counted. And miraculously, Pythagoras was correct.

The fishermen were amazed. They asked Pythagoras what they should do, and his only instructions were to return the fish to the sea. It's easy to imagine the debate that must have ensued over whether or not to obey. What unimaginable grimaces must have crept over their faces as they released a day's livelihood back into the drink thanks to the sage's freakishly lucky guess. But as hard as it must have been, the fishermen let the fish go. And according to the story, all the fish survived and swam away, even though they had been out of water for some time.

Then Pythagoras showered appreciation upon the fishermen. He opened his purse and paid them a fair price for the fish. The fishermen's frowns must have turned into the brightest smiles at this, and sometime later they probably told everyone they met about their strange encounter, helping to foster the legend of Pythagoras.

There is no way to tell if this really happened. The stories of Pythagoras the mythical figure were so exaggerated that it is difficult to really know what to think about Pythagoras the man. One twentieth-century authority says of him that of the man we know little and of his writings, nothing. But he succeeded in leaving his mark on mathematics thanks in part to this reputation.

Pythagoras was one of the most successful teachers of his day— perhaps one of the most successful of all time. He built a school that attracted masses of people and persisted for centuries. Moreover, he had an inner circle of monklike students who were so devoted to his teachings that they gave up all their possessions to live with him and spent their days practicing, memorizing, and reciting his teachings. These students went well beyond what one would expect from any devoted protégés. They did not just praise their master's virtues. They credited him with their own discoveries. Thus Pythagoras the man was really more than one man or woman. He was a movement. And one of his and his followers' greatest interests was in geometry, which he probably learned in Egypt, where it was used for thousands of years before his time. Because of this, Pythagoras became one of the most important mathematicians in history, even though, ironically, he may not have been such a great mathematician himself.

Herodotus placed the origins of geometry as far back as 3000 BC in Egypt, but it is impossible to say for certain when the Egyptians invented this branch of mathematics. It was so common in Pythagoras's day that Egypt was said to have "bristled" with geometry.

Egypt was an anomalous land. Its location in the middle of a massive desert made it safe from invasion. Any threatening army would have to cross many miles of the most inhospitable terrain imaginable. This freedom allowed Egyptian society to advance, and the Nile made life an oasis in the middle of this scorched earth. This mighty river wended its way down from the runoff-rich headwaters in modern-day Ethiopia and Rwanda. And in the middle of the Egyptian desert the river dropped a load of rich silt with every flood.

In order to survive as a civilization in the middle of a desert, the Egyptians had to maximize the gift of the Nile each year, and so they invented the technologies they needed along the way. One of their innovations was a useful calendar. Because their way of life depended on the annual floods, knowing approximately when the flooding season would begin was a matter of survival. So they began to look to the stars for guidance. They designated July as the beginning of the year, indicated by the rise of Sirius, the brightest star in the sky. When the bright star rose, people knew to get ready for the floods. This calendar was the basis for the later Julian and Gregorian calendars.

The Egyptians also invented geometry, the branch of mathematics that deals with the relationships of points, lines, angles, surfaces, and solids. It was useful for when the floodwaters receded, as a practical set of tools for surveying land. The very word *geometry* comes from the Greek words *ge*, which means "the earth," and *metreo*, which means "to measure."

The rising and falling of the Nile obliterated the boundaries between plots of land and compelled the Egyptians to apply geometry to measure, divide, and ultimately cultivate the delicate river watershed and floodplains upon which their agriculture wholly depended. They stretched ropes between various points to establish boundaries. People would be taxed on the basis of the area of land they farmed, but every year that area would change. Give an Egyptian a piece of rope, and he could measure out a connecting piece, marching off at a particular angle so that a third piece of rope, enclosing the area of land into a triangle, would be of an exact reckoning. Over the centuries, the Egyptians measured the earth well and painted their corner of the desert with color and life and created works of art and architecture such as the world had never known.

All this is revealed in an ancient papyrus written about 1650 BC by a scribe named Ahmes. Titled "Direction for Obtaining Knowledge of All Dark Things," the paper contains eighty-seven practical mathematical problems, including how to divide a given number of loaves of bread among different numbers of individuals, and how to measure the area of a field.

The Egyptians were not alone in their mathematical progress. The Babylonians also had a strong tradition of mathematics, which they also developed out of necessity. For them, it was a practical tool for urban planning and commerce. Merchants need math skills as much as carpenters need sharp tools.

The Babylonians knew about certain geometrical theorems, and they made amazing progress in arithmetic. They worked with squares, square roots, cubes, cube roots, and exponential functions for computing compound interest. Tables of these numbers still exist. The Babylonians divided the circle into 360 degrees and the day into 12 hours of light and 12 hours of darkness. They made an hour 60 minutes and a minute 60 seconds. They employed cuneiform texts that reveal they understood the relationship that came to be called the Pythagorean theorem.

Science and mathematics were once more like religion—or perhaps in today's highly charged atmosphere of attacks on public teaching of evolution it would be better to say that they were once unlike what they are today. Back then, thousands of years ago, they were trade secrets. Mathematics was probably a collection of practical bits of rote knowledge that was taught and memorized by generations of teachers and students without proof and with an eye to splitting loaves of bread and dividing plots of land. It is perhaps misleading to even think of it as mathematics, since it was not the rigorous, fully examined system we call by that name today. It was much more practical and boring—something like fishing.

In the mathematics of the ancient, you bait the hook, you lower it into the water, you wait, and then you pull. There was no higher examination of how to bait the hook or how to use the basic technique of fishing to sample the population and distribution of sizes of fish in the Nile. Mathematics was simply bait and pull.

The Egyptians and the Babylonians weren't interested in theory for its own sake but in the practical mathematics that worked. Both cultures had their mathematical difficulties as well. The Babylonians used an unwieldy base-60 system of numbers instead of our more familiar base 10. The Ahmes Papyrus reveals that multiplying 41 by 59 required a complicated eleven-step addition.

Egyptian arithmetic was so rudimentary, in fact, that some modern writers have been cool or even cruel in their assessment of its value. Some have even questioned whether the Egyptians contributed anything at all to the development of mathematics. Sure, they had mathematics, but they had no systematic approach to the subject. *They had no proof.*

Geometry stagnated under the Egyptians. They never really advanced the subject beyond the practical art of rope stretching. Another Egyptian treatise written some two thousand years after the Ahmes Papyrus showed no improvement from the more ancient ways. The Egyptians had no system of postulates and logically deduced proofs. They took no systematic approach to it at all. Instead, mathematics was, as Ahmes said in his papyrus, a set of directions for figuring out "dark things."

But these *were* practical things, after all. And people from other cultures had plenty of use for them. As often happens with such technologies, the mathematical secrets leaked out. That is how the Greeks came into the picture. Under the Greeks, mathematics would advance much further and evolve into a rich and wonderfully complicated subject. Geometry was still practical, but it became an art as much as a science. So if there is anything one can say about Egypt and its contribution to Western mathematics, perhaps it is this: they were leaky with their secrets.

Pythagoras was crucial in this transition, but he was not the first person to discover geometry in Egypt. He was preceded by untold others who lived, as he did, in border cities where rich cross-cultural exchange was possible with countries like Egypt. One of these mathematicians was Thales, who was slightly older that Pythagoras and is thought to be the oldest of the seven sages of ancient Greece. He was also one of the greatest mathematicians and one of the first to learn geometry from the Egyptians. Like them, Thales's interests seemed to be primarily in practical application, but unlike the Egyptians, he moved it further. Thales founded abstract geometry. He gave the first general explanation of the universe, and he was perhaps the first person to concern himself with proving theorems. He did this after visiting Egypt.

Little is known about Thales with certainty. Many of his biographers lived seven hundred to one thousand years after he died. They were not people who shared a time or even a common culture with him. Nor do they agree on all the aspects of his life. He may have been Miletian or Phoenecian. We cannot know for sure.

That did not stop many interesting stories about him from accumulating through the years. Some of these portray him as an addled scientist, lost in his own mind. According to Plato, Thales was on one occasion so concerned with his studies of the stars that he stumbled into an open well because his gaze was fixed on the heavens as he walked.

Most of the stories show how Thales was a wise man of the first order. Herodotus claimed that Thales predicted the eclipse of May 28 in 585 BC. Aristotle noted that he was a savvy businessman—perhaps the first mathematician who took an analytic approach to commerce in the sense that he predicted a short-term trend and sought to gain economic advantage by applying it to the market.

Thales predicted a bumper crop of olives one year. It was going to be a particularly good harvest, he thought, and so he bought up all the olive presses that he could with the intention of renting them out at a steep markup when it came time to squeeze out the olive oil. And he was absolutely right. It was a bumper crop, and he made a fortune from his neat monopoly. Then at some point, he went to Egypt and learned geometry.

There is no proof of any of his adventures in Egypt, but the legend survived through the ages, if for no better reason than it makes such a good story. More than that, it makes sense. According to Plutarch, Thales was a merchant and this allowed him to visit Egypt.

In going to Egypt and learning geometry, Thales acquired a great practical tool. He applied it to the sort of problems that were relevant to someone living in a seaside community. He used geometry to come up with a way of measuring the distance from a seafaring ship on the horizon to the shore—significant in his seafaring culture for knowing when to be at the docks ready to greet the boat. Thales also measured the height of the great pyramid at Giza, which was already some two thousand years old when he was alive. It had been around so long that nobody knew its height. There was no easy way to measure it.

Obviously, climbing the pyramid and directly measuring its height with a tape measure wouldn't work because the pyramid itself would get in the way. The best you could do would be to measure one of its sides. You could measure the height directly by erecting an enormous platform next to the pyramid, but structurally there would have been no safe way to build a platform so high—not to mention the overwhelming amount of time, labor, and resources needed to create such an absurd construction (the eighth wonder of the ancient world: a scaffold tall enough to measure the seventh).

Thales simply observed the shadow the pyramid cast upon the desert sands and placed a stick in the ground just at the edge of the pyramid's shadow. He chose a time of day when there was a long shadow—probably late in the afternoon or early in the morning when the stick's shadow was the same length as the stick itself. At that moment he could measure the height of the pyramid through a simple trick: the shadow measured the pyramid's height. The distance from the center of the pyramid to the stick, which was easily measured, was the exact height of the pyramid.

The brilliance of this was that he solved the problem mathematically by using the equivalence of similar triangles—that two similar triangles will have the same ratio of sides. Thus at the exact moment when the length of the stick was the same as the length of the stick's shadow, the length of the pyramid's shadow was the same as its height. It was a simple but crucial concept.

Back in Greece, Thales started a school. He is a remarkable thinker because he was the first to come up with an overarching philosophy to explain the material world (he said all is water). Pythagoras visited him in this school as a young man. Thales had a profound effect on the course of the young man's life because he encouraged Pythagoras to travel to Egypt, just as Thales himself had done many years before.

An interesting thing happened to Pythagoras on his way to Egypt. He hitched a ride on a boat heading to the Middle East. Ships frequently came and went, and berths like these must have been common in his days. No doubt robberies were as well, since traffic in contraband would have been the same to sailors as traffic in traded goods: profitable. Stealing things from their passengers was not the worst thing that nefarious sailors could do, either.

In Pythagoras's case, since he was a young man, the sailors on his boat must have decided that the most valuable possession on his person *was* his person. He was young, fit, appeared to have a good head, and certainly had a nice face. So they decided to sell him into slavery once they reached port. Pythagoras was no fool, and he may have been aware of their plot. His actions, though, were not those of a man worried about the yoke. He behaved as though lost in deep thought, with concern for much greater things than his immediate safety. Was this showing obliviousness? A lack of concern for his own safety? Or did he come up with the most brilliant way to defeat the slavery plot: sit there and do nothing?

As the boat made its way along the Mediterranean coastline, bobbing in surf, sitting still was the only action Pythagoras took. And as the days and nights peeled away, he continued to sit there and do nothing. At first, sailing with Pythagoras on that small boat must have been strange for the sailors, odd bird that he was. He was more of a statue than a man. The sailors observed him sitting there perfectly still for days, neither eating nor moving, until finally they had a change of heart. Their initial curiosity turned into fascination and then into respect, and by the time they arrived on the Egyptian coast, Pythagoras's destination, they had formed a near reverence for this man. When they struck land, they lifted young Pythagoras aloft and carried him ashore. There they built an altar before him, loaded it up with fresh fruit, bade him farewell, and returned to their ship.

Nobody knows what happened to these sailors next. Perhaps they amassed some small fortune and retired into one of the seaside cities of the ancient world, living out their days in pickled contentment. Perhaps they died the violent, forgotten deaths of lost sailors cut down on the tips of rocks or swords or choking silently in open water. Or perhaps they escaped into a happier though equally obscure fate. No doubt, wherever they wound up, they always recalled their strange statuesque passenger. Pythagoras, on the other hand, went on to lead one of the most memorable careers a mathematician could hope for—long, diverse, meaningful, and with a lasting impact on humanity.

Pythagoras was born around 580 BC on the ancient island city of Samos, off today's Turkish coast. He grew up in a culture that was thriving at the dawn of Greek civilization in the sixth century BC.

Mnesarchus, Pythagoras's father, an engraver by trade, was away on business when his wife, Pythias, discovered that she was pregnant. Mnesarchus was in Delphi, where the famous oracle prognosticated on the fates that awaited those who came with offerings to the temple. According to legend, the oracle told Mnesarchus that his wife was pregnant and that the baby would be an ornament to humanity. The oracle promised Mnesarchus that his son would "surpass all who had ever lived in beauty and wisdom," and that he would "be of the greatest benefit to the human race in everything pertaining to human achievements."

In a sense the prophecy may have become self-fulfilling, because how could any expectant father hear this and not be affected? The prophecy must have raised the baby all the more in his estimation, and Mnesarchus would apply all the sweat and care spent on a lifetime of engravings to shape Pythagoras's person. Likewise Pythagoras's mother, who was the most beautiful woman on the island of Samos, must have lavished attention on the child, though it is doubtful that she would have needed the oracle's promise of Pythagoras's future greatness to dote on her son—what mother ever needs an excuse for that?

Pythagoras enjoyed these advantages, and they paid off. He studied under some of the greatest teachers of his day. There was Pherecydes of Syria and Hermadamas, the son of the famous Creophylus. He may even have studied under Thales and Anaximander, who is famous for studying astronomy, biology, and geology, and for his belief in the boundless nature of the universe. Even as a young man, Pythagoras's reputation for wisdom spread throughout the cities of the ancient world and preceded him when he traveled, which he did extensively.

In the Near East Pythagoras lived among Arabian and Jewish communities. He also lived in Babylon and spent a long time in Egypt, where he visited every wise man, holy site, and temple that he could. Then, having learned all the mathematical secrets he could, Pythagoras traveled back across the Mediterranean and, finding himself unable to live under the local tyrant Polycrates, left Samos and settled in the city of Croton in the Calabria region of southern Italy where he established his popular school around 530 BC. He was over sixty at the time.

Pythagoras's arrival in Italy was a key step in the history of geometry because of the way he pushed mathematics on his followers. It was due to him as much as anyone else that the discipline took root and flourished outside of Egypt. He would himself disseminate all that he knew on the subject, but more important, his influence would promote its further study, development, and advancement. Mathematical theorems were like divine praise to him. He was enamored by the number 10, for instance, because it is the sum of the first four integers $(1 + 2 + 3 + 4 = 10)$.

Geometry in Greece underwent a significant transition after Pythagoras. Greek intellectuals who followed him embraced it, and the subject grew from a practical art to be employed toward some application into a subject to be studied for its own sake—for the love of learning. Pythagoras's followers, the Pythagoreans, discovered irrational numbers. They proved that the sum of the angles of any triangle equaled 180 degrees. (This was something that would later be cast aside by non-Euclidean geometry.)

Few mathematicians in history have had a more devoted group of followers. They were a secretive cult. According to legend, the first person in the order who supposedly divulged the secret of the irrational number $\sqrt{2}$ to outsiders was shipwrecked by the gods themselves. The Pythagoreans emulated Pythagoras's actions and repeated, memorized, and obeyed his speeches. They attributed to him their own discoveries and inventions, as well as amazing, even supernatural, powers. In one story, Pythagoras had the ability to appear in several places simultaneously. In another, he made the dire prediction that a ship seen sailing into the city's harbor would be discovered to be full of dead people (it was). One time when he was enjoying the Olympic games, eagles supposedly swooped down to him from Mount Olympus. They landed next to him and allowed him to pat them before soaring off again. Another time, Pythagoras approached a wild bear that was terrorizing a neighborhood. Like an ancient Grizzly Adams, he reasoned with the animal in bear language to leave the poor humans alone, and he gave the bear a meal of grains and fruit. They parted ways, the bear retiring into peaceful obscurity.

Pythagoras was also an ox whisperer, which he proved when he came across one of these beasts helping a farmer harvest a field of beans. He whispered in the ox's ear and convinced him to give up eating beans. Not only did the ox do as Pythagoras requested, swearing off beans for the rest of its long life, but it turned over new leaf, becoming a sort of holy four-legged sideshow attraction at the temple to Hera. So worthy was Pythagoras's reputation that a Roman biographer named Porphyry said of him centuries later, "Never was more attributed to any man nor was any more eminent."

Never was a historical figure more difficult to assess, either. What we know about Pythagoras, we question. He left no book behind with his teachings, and in the centuries after his death, his accomplishments were so exaggerated that it's difficult to know exactly what he did or said. Confusing the situation even more, perhaps, is that there were several people who shared his name in his day. There was Pythagoras the wrestling coach. There was Pythagoras from Zacynthus. There was Pythagoras the sculptor who lived in Rhodes. There was a sculptor in Samos names Pythagoras, a Pythagoras who was a public orator, and another who was a doctor who dabbled in writing. Finally there was a historian named Pythagoras. All of these men lived at about the same time.

When the most famous Pythagoras started his school, it became quite popular, and he attracted throngs of admirers. Crowds would flock to hear him, and, whether fact or legend, he once made a speech that attracted an audience of two thousand to his school by the time he was finished—many more adherents than some of history's greatest prophets garnered over their entire lifetimes. His school paid personal dividends as well: he married an extremely attractive, intelligent young woman who was an admirer of his teachings. What were these teachings like?

Pythagoras's school sounds more like a cult than anything else in our experience today—a cult based on a strange mix of pseudo-religious dictums, dietary restrictions, and daily rituals that make the Pythagoreans seem perhaps weirder than they actually were. But this merely reflected Pythagoras's far-flung interests in everything from moral behavior to music, medicine, and mathematics.

It is not known whether it was Pythagoras or his disciples who invented music theory. He originated the idea that musical intervals can be expressed as numerical intervals. He discovered that a string stopped at half its length rings an octave higher and that at two-thirds its length it is a fifth higher. Modern music theory and instrument design derives from this simple observation. So influential was Pythagoras's work in music that centuries later Plato considered music theory equal in importance to astronomy.

Pythagoras had a close inner circle of disciples who would sit with him and gain the deepest pearls of wisdom from their master. He also had a much larger outer circle of followers who were physically kept at a distance from him and allowed to hear his teachings only in the form of riddlelike aphorisms with no explanation of their deeper meanings.

Some of his teachings were simple proclamations on subjects ranging from how to behave in public to cosmology to obscure bits of detail, such as the type of wood that coffins should be made out of. Others were mini-lessons in the form of questions and answers: What is the wisest human thing? (medicine). What is the most just thing? (to sacrifice). What is the most beautiful? (harmony). Some of the teachings were straightforward, such as recounting the day's activities while lying in bed at night and then thinking of the coming day's actions in the morning immediately upon waking. Others seem quite insightful, especially for their day. For instance, he preached friendship and advised his followers to come to terms with one's loved ones and remove rivalries and scars from friendship.

Still other of his teachings seem arbitrary, even strange—such as the dietary restrictions Pythagoras placed on himself and his followers. He avoided wine and meat, and he absolutely forbade the eating of hearts. One of his most famous dietary restrictions was on beans or lentils. In fact, all foods that caused flatulence were forbidden. Instead, Pythagoras ate simple foods like honey for breakfast and grains and herbs for dinner. Initiates to his inner circle had to follow these restrictions closely.

Put simply, Pythagoras was as much a lifestyle as a man. He and his followers lived as minimalists, shunning any form of luxury and the notion of common property. They also had to give up their worldly possessions, but this was a forgiving form of self-sacrifice, because if it didn't work out and the person left the fold, he got twice the value of his goods back.

One of the strangest practices that the Pythagoreans followed was their initiation ritual. Someone could not just walk into the group and give up all his possessions. He was subjected to a rigorous mental, behavioral, psychological, and physical exam. His ability to retain information and repeat it back was tested. His passions and his responses to various psychological stimuli, as we would call them today, were examined. So were his actions and temperaments. During this time of testing, a person had to observe complete silence for years, quietly listening to Pythagoras's teachings. But he was forced to do so from a distance, never being allowed into the same room as Pythagoras—not even to see him for a moment.

When a person finally was accepted, his life from that point on would resemble a monk's. The Pythagoreans lived communally. They took their meals together. They exercised together; they read and studied together; they looked into the mysteries of math, life, and the universe together. In the morning, they would go for walks by themselves in quiet places, reflecting. Then they would exercise, racing or wrestling. After, they would dine on simple foods like bread and honey. Then they would converse, take walks in groups, practice their lessons, wash, eat a simple supper, and enjoy more conversation.

It sounds so quaint in some ways—even more so when you envision the Pythagoreans in their period dress, bare flesh spilling out of white robes, sitting around, discourse and laughter filling the nighttime air, applying poultices to one another and testing one another's memorization skills. Here is a course of grains and flavorful herbs. There is Pythagoras strumming on his stringed instruments—one can imagine he was quite good. Here they all are dancing around. Poetic verses. More strumming. A sea of unkempt hair on every head swaying as they bob to the plucking of Pythagoras.

Pythagoras, abstracted by veils himself, was the first person to really put abstract reasoning into mathematics. He set geometry on a path from which it would never turn back. The art of geometry would survive long after he and the Pythagoreans were gone.

In the end, it is impossible to know what to attribute to Pythagoras or his school. The Pythagoreans were not around long enough for people to really understand who they were or what they did. The accounts of them are not firsthand and were often written centuries after they had disappeared. Besides that, the Pythagoreans were at heart a mysterious cult that upheld a common pact not to disclose the revelation of mysteries that bound them together.

Ironically, as strong and cohesive as this sort of exclusivity may have made the Pythagoreans, it was ultimately the cause of Pythagoras's demise. His end came about as a sort of backlash against him and his cult. Popular opinion of him being what it was—worshipful—there were bound to be those who wanted in to his inner circle. And there were bound to be those who would try and fail to breach it, including some who would take rejection as free leave to detest Pythagoras and the Pythagoreans and would plot against them.

One such plotter was a wealthy and powerful man named Cylon, who was from Croton, where Pythagoras had established his school.

Cylon was of the most privileged class, but despite the many advantages he no doubt enjoyed in life, descriptions of him make him sound like a cruel and evil man. His disposition, for instance, is described as severe, violent, and tyrannical. But because he was a child of the most superlative privilege, Cylon saw it as his right to be accepted into the Pythagorean fold, and he sought out Pythagoras and his companions, probably expecting that they would be as open to him as any other privilege he might seek—simply a matter for his convenience and choosing.

But joining the Pythagoreans was not Cylon's choice to make. Pythagoras, being given to extolling the virtues of moderation in his teachings and living by the most rigorous asceticism himself, gave no special consideration to Cylon. Spurned, Cylon started a conspiracy to oppose Pythagoras, which led to what have been called the Pythagorean riots.

There are multiple accounts of what happened next, and all end in some form of tragedy. These stories are rich in detail, much of which may be false, but they add to the myth of Pythagoras the god. Like many great legends, he did not die a quiet death in old-age obscurity, but was tested in the fires and swords of those who despised him.

The plotters succeeded in stirring up enough resentment toward Pythagoras that one hot evening (as it might be imagined), they grabbed the opportunity to finish the cult once and for all. That night, a group of Pythagoreans were cornered in the house of a man named Milo, a wrestler. The mob set upon the house with stones and torches, burning it and stoning the Pythagoreans inside.

By one account, all inside died except for two men. By another account, a group of Pythagoreans escaped only to be pursued by the mob and cornered at the edge of a bean field. Further flight was cut off—except over the beans—but that would require them to touch the beans. Apparently Pythagoras's statement not to touch beans was a principle so strictly adhered to that his disciples would not cross the field. Instead they chose certain death at the hands of the angry mob. They turned and fought with their meager sticks and fists against superior numbers and weaponry, and were killed.

It was a loss for the ages. "When died the Pythagoreans," wrote the philosopher Porphyry in his *Life of Pythagoras* at the end of the third century, "with them also died their knowledge, which till then they had kept secret, except for a few obscure things which were commonly repeated by those who did not understand them." The fate of Pythagoras himself is disputed by this and various other writers, none of whom were actually present at the conflagration to begin with. Some say that Pythagoras was not present that night. Others have him there, witnessing the blaze and escaping among its survivors. In one account, his followers threw themselves into the flames, forming a human bridge so that Pythagoras could step to safety across their scorched bodies.

Once free of the burning building, Pythagoras was chased from town to town by the crowds until he finally found refuge at a temple where, according to one story, he slowly starved to death over the course of several weeks. Another account says he died from grief, not lack of food. Yet another version has Pythagoras leading a contingent of followers on the escape that came to a sudden halt and resulted in their bloody end at the edge of the bean field.

Pythagoras may have been eighty when he died, perhaps even ninety. By one account, he was 104. But whatever the timing of his death, regardless of whether he died in some faraway temple or at the edge of a bean field, or whether he was a living god or a refugee, Pythagoras occupies a unique place in the history of mathematics.

The most obvious Pythagorean legacy today is the mathematical formula that still bears his name. The Pythagorean theorem is perhaps the most applied theorem in all of mathematics—certainly in elementary math—and also the most famous (although in recent years Fermat's last theorem has given it a run for its money). It has been proven many ways through the years. The Pythagorean theorem is easily memorized as "A squared plus B squared equals C squared," equating the sides of a right angle with the length of the hypotenuse that intersects them.

If we are to believe Pythagorean legend, Pythagoras was so excited when he discovered his theorem that he immediately sacrificed one hundred oxen. As one writer put it, "When the great Samian sage his noble problem found, a hundred oxen with their life-blood dyed the ground."

Like a marble temple from the ancient world that's still standing, the Pythagorean theorem has enjoyed great longevity, staying as remarkably simple today as it was twenty-five hundred years ago years when it was discovered. It is one of the most often-repeated mathematical formulae and is still taught to nearly every schoolchild in the world today.

Ironically, as great an achievement as the theorem was, Pythagoras may not have been the one who discovered it. One of his followers might have been the real originator, later attributing it to Pythagoras out of respect for the master. Or perhaps the formula was something that Pythagoras carried back from Egypt, where the theorem may have already been known in some form when he was there. The identity of certain triangles that today would be called "Pythagorean triples" was certainly already known. These are special right triangles whose sides are measurable in simple whole numbers. The simplest of these, the right triangle with sides measured in units of 3, 4, and 5, can be found among the other problems in the papyrus written by the scribe Ahmes many years before Pythagoras was born.

In any case, Pythagoras left a much greater legacy in the deductive reasoning he helped to introduce and his followers helped to formalize. Following Pythagoras, his followers worked to improve mathematics, transforming it into a subject of truth rather than experience. Pythagoras also gave birth to the idea of mathematical harmony. It was a compelling idea—so much so that it survives in some form today.

"Pythagoras transformed mathematical philosophy into a scheme of liberal education, surveying its principles from the highest downwards and investigating its theorems in an immaterial and intellectual manner," Eudemus of Rhodes wrote. His strongest asset seems to have been his ability to influence people, and in this respect, Pythagoras the god and Pythagoras the man were the same. In the end, it really doesn't matter what he did or didn't do mathematically. It was his influence more than his work that counted. Pythagoras was the great popularizer. The theory of numbers as it was taught in his school became book seven of the *Elements.*

If you like to think in basketball terms, Pythagoras was no Michael Jordan. He was not the gifted prodigy who would define the game in his lifetime and inspire generations of players who came after him. Pythagoras was more like a beloved basketball commentator—a broadcasting legend whose love of the game would actually improve it. He set the stage for the love and practice of mathematics in the classical world, which after him exploded.

Invoking the name any sports figure, however, is too narrow to adequately describe Pythagoras. His legend has so many dimensions that comparing him to any one person famous for any single extraordinary thing does not do him justice. He was a constellation rather than a single star—part man, part myth, and only visible through the imagination.

After Pythagoras died, his brotherhood scattered. The center of learning shifted from southern Italy to Greece. His greatest legacy, that deep appreciation for mathematics, persisted long past the Pythagorean riots. He planted the love for geometry in fertile ground, and after him this love flowered. The Greeks turned geometry into an almost mystical and sublime art, and for four hundred years they developed it not for practical application but for mathematical truth. Where the Babylonians and Egyptians may have known and used many different geometrical identities, they did not bring them into a logical system of proof, which is something for which the Greeks will always be remembered. Geometry was Greek in name and spirit—logical, analytical, and so very useful.

The great cultural expansion of the Greeks came after the classical period following the end of the Persian wars, from 500 to 480 BC. Around 450 BC everything began coming together intellectually in Greek culture, and for the next two centuries more great discoveries and works of art were created than perhaps at any other time in history. The fifth century BC saw an incredible parade of geniuses—Zeno, Anaxagoras, and Democritus to name just a few. Their contributions to science, mathematics, and mechanics continue to resonate today.

Geometry flourished in Greece through the age of heroes, the Persian and the Peloponnesian wars, the victories and defeats of Athens, and the lives and deaths of people like Pericles, Socrates, Plato, and the great historian Thucydides, whose history of the Peloponnesian war is a literary classic. It was also a time of great creative expression in art, literature, science, and philosophy.

Some of the mathematicians of those golden days of the fifth century BC are really known only by name, not by their works, which have been lost. Their names are remembered because books that survived the ancient world refer to them—names like Mamercus, brother of Stesichorus the poet. Hippias of Elis, Oenopides of Chios, Theodorus of Cyrene, Leodamas of Thasos, Archytas of Tarentum, Theaetetus of Athens, Neoclides, and Leon. Names and names and a few references to their discoveries are all we have of this ancient pantheon of minds who for hundreds of years created and advanced geometry.

This same century was also the one in which lived one of the greatest mathematicians of the ancient world, Hippocrates of Chios, who dwelled in Athens around 440 BC—not to be confused with Hippocrates of Cos, who also lived in the fifth century BC and is considered the father of medicine, the man for whom the Hippocratic oath is named. Hippocrates of Chios wrote the very first textbook of geometry. He pioneered the logical deduction of proof and invented the method of proving an idea by showing its contrary is ridiculous—something now called a *reductio ad absurdum* ("reduction to the absurd").

By 350 BC Greek civilization was at its apex. From there it penetrated deep into the rest of the world, carried on the backs of Alexander's armies as they marched into the heart of Asia. During this time, geometry and other areas of mathematics were collected, corrected, and codified by a succession of thinkers. Theaetetus of Athens discovered some of the work around 380 BC—especially solid geometry. Some of his efforts became book ten of Euclid's work. Eudoxus of Cnidos worked on material that found its way into books five and six of the *Elements*. Theudios of Magnesia, Dinostratos, and Menaechmos all did great mathematical work at Plato's academy in the mid-fourth century BC. They explored the elements of geometry, wrote a textbook on the subject, and taught it to many others According to the Roman writer Proclus, this book "put together the elements [of geometry] admirably, making many partial propositions more general."

There was by this time a definitive history of geometry, written just before Euclid's time by a Greek mathematician named Eudemus of Rhodes. Eudemus was a student of Aristotle's and compiled his book in the late fourth century BC. It became a standard reference for hundreds of years before it was unfortunately lost. All that remains are a few scattered passages that later writers saw fit to quote.

Generation after generation of Greek mathematicians invented the basic methods of modern mathematics piecemeal. They made definitions and mastered ways of proposing and proving mathematical truths—called postulates—based on a few starting points. By the time the conquest of ancient Greek cities under Philip of Macedon and his son Alexander the Great began in the fourth century BC, numerous masterworks existed that offered systematic treatments of the mathematics of the day. In one way, the decline of Greece's supremacy in intellectual matters followed its defeat by the armies of Philip of Macedon. Greece's flower was wilting.

But in another way, the advance of Greek mathematics really owes as much to Philip and Alexander as it does to the nameless and famous mathematicians who advanced it in content, because they encouraged its teaching and spread. Alexander even included some scientific research in his military campaigns—an idea no doubt inspired by his teacher Aristotle, who was the first to consider organized research and the first to classify knowledge into the different disciplines that existed in his time. However crude the science that followed Alexander's brutal conquest of the ancient world, they were the first scientific expeditions.

3

The Mystery Maker

The first and last thing required of genius is the love of truth.

—Goethe

One of the greatest scientific expeditions through the centuries was an imaginary one—the mental journey, which many mathematicians made, to solve the fifth postulate. Carl Gauss made this journey himself as a young man, and throughout his lifetime, he was friends with several other people who did as well. While Gauss was a student at the University of Göttingen, one of his best friends was a fellow traveler on this road.

Farkas Bolyai von Bolya was a talented mathematician, artist, author, musician, and scion of a famous old Hungarian family. He is still remembered in Hungary for his plays, but more significant than his contributions to literature was his life in mathematics. He enjoyed a distinguished career and lived a dignified life worthy of his family's name, such as it was. His family had lost its fortune over the years, and by the time Farkas was born, there was little left. Farkas was not alone in this. He belonged to a whole class of once fabulous but now impoverished nobles—mockingly called the "sandaled gentry" because they could not afford boots.

Farkas was slightly older than Gauss but a year behind him in college, and the two almost did not meet. Farkas had been hired by a local baron to be a study companion to his son, and he had been distracted on his way to college by a visit to Vienna. He so loved the city and the artillery school he visited there that he came close to dropping his plans to enroll in the University of Göttingen, determined to become a professional soldier instead. Only an errant spark convinced him to abandon these plans. He had almost

45

blinded himself in an accident involving some homemade gunpowder and that spark. One explosion was all he needed to convince himself not to become a soldier.

Instead of pursuing a career in the army, Farkas went to college at Göttingen, and he soon met Gauss at the house of one of their professors. Gauss saw in Farkas a kindred spirit and fellow mathematical enthusiast, someone with whom he could share his innermost mathematical insights. They became fast friends—brothers of sorts who shared a common passion. Gauss was so fond of his friend that he gave Farkas the tablet on which he had worked out how to construct his famous seventeen-sided polygon. He once commented that Farkas was the only one who understood his views on the fundamentals of mathematics. Other mathematicians may have been able to discuss ideas with him, but Farkas was one of the few who could really follow him.

Gauss was a mathematical genius, so this says a lot about Farkas. He had to be one of the sharpest minds in town to be able to keep pace with Gauss. On the other hand, this may not have been completely obvious to everyone who met Gauss. Farkas once said of Gauss that he was so humble, one could know him for years without recognizing how great a genius he was.

Indeed, Gauss was a quiet student while at Göttingen. Farkas said that they used to spend many hours together, often going on long walks during which they would often discuss mathematics, but frequently they would remain silent, resting in their own thoughts and sharing the "invisible passion for mathematics," as Farkas put it. Once they walked about 140 miles, all the way from Göttingen to Brunswick and back, to visit Gauss's mother. Still, Farkas later regretted this silence. "It is a pity that I did not know how to open this silent, untitled book and read it," he said one time.

One of the reasons for Farkas's regret was that he and Gauss never helped each other solve one of Farkas's main passions: proving the fifth postulate. It was one of the burning mathematical questions of his day, and Farkas sought the solution to it for most of his life, blindly chasing it for so long that he later regretted wasting so much time on it. It became his white whale. In his autobiography, Farkas wrote, "I drove this problem to the point which it robbed my rest, deprived me of tranquility."

Gauss didn't waste his time trying to prove the postulate, though he certainly discussed the mystery with Farkas. They both had learned geometry the way that it had been taught for more than two thousand years. The subject had been adapted, revised, and updated, but it was still more or less true to its original source— from the *Elements*, perhaps the greatest mathematical book of all time.

After Alexander the Great conquered Greece, he founded the city of Alexandria in Egypt, in 332 BC. Two years later, his general Ptolemy Soter sought to transfer the flower of Greek intellectual pursuits there. Almost from the time its foundation stones were set, so was the city's fate. Within a generation it supplanted Athens as the greatest center of intellectual activity in the ancient world. The time of Egypt had arrived.

The culture of Alexandria was not solely Egyptian, of course. It was primarily Greek. And so the intellectual pursuits were Greek as well. A large school of mathematics grew there, and Greek scholars brought Greek-adapted mathematics, ironically returning geometry to its long-lost native soil. Alexandria was built on a narrow ridge of limestone separating the Mediterranean Sea from a great lake called Mareotis. This ridge of land separating the two bodies of water also sandwiched a great channel of water. Offshore, a long island called Pharos broke the crashing waves and provided a safe port. This port was also clean—far enough away from the swampy dregs where the Nile finally emptied into the sea after its long journey from the heart of Africa. It was a gorgeous spot, boasting beautiful weather, lots of freshwater, and accessibility by riverboat to the heart of Africa and by ship to the whole of the Mediterranean. Alexandria was the perfect place to contemplate higher truths and mathematics.

Alexander had once gazed upon the empty land that would become Alexandria and seen a great city lying there. Could he have known that within a generation it would sparkle with genius and become an icon to the greatness of his culture, which then dominated the known world? Could he have known that this gem would be at once an emblem of the greatness of Egypt's past and a crowning jewel in his own empire?

After Alexander died, this empire was split. Alexandria and the rest of Egypt would be ruled by Ptolemy Soter, who reigned from 323 BC to 285 BC. Ptolemy took his charge very seriously. Alexandria was not just one of the jewels of the empire. For him it represented power and wealth. He was a no robber baron, though. He sought to be a beneficent ruler and poured resources into his city.

He opened up a large school there inspired by the garden where Plato and his disciples used to meet. Plato's followers called this garden the "the academy," because the land where it stood was owned by a man named Academos. In our modern, academia-influenced world, it hardly seems possible to consider the word divorced from its meaning, but back then it signified a certain place. Ptolemy Soter's importation of the idea of the academy was the beginning of the use of the term to describe the collegial open exchange of ideas between scholars engaged in searching for higher knowledge. Just a few centuries later, Cicero celebrated this tradition and the exchange of ideas by naming his country house "the Academy."

One of the most important sets of ideas exchanged in Plato's academy concerned mathematics. Above the entrance were the words "Let no one unacquainted with geometry enter here." Plato may have been inspired by the Pythagoreans and their school in this regard, and by keeping the mathematical tradition alive, he ensured it would survive for centuries more. This was a tradition that Ptolemy fostered and improved when he established his school and library at Alexandria.

Libraries were nothing new when Alexandria began its famous one. The Egyptians, Phoenecians, and Chaldeans all had large libraries to keep archives of their records. But the world had never seen anything like the one established at the tip of Africa. Ptolemy contributed tens of thousands of manuscripts to it, and his successors gave even more. By law, any books that were brought into the city had to be submitted to the library and copied if no version already existed there. Merchants who hailed from Alexandria were asked to purchase manuscripts whenever possible.

As famous as the library is today, the city of Alexandria was equally famous in its day for its community of scholars. Ptolemy brought together poets, philosophers, mathematicians, artists, physicians, astrologers, historians, and writers. Alexandria had a university, the academy, a library, and a museum. One of the revolutionary changes Ptolemy instituted was to pay his scholars a salary and give them room and board so that they could pursue their studies. This also allowed the city to attract a great number of outstanding minds.

One famous scholar was Archimedes, the greatest mathematician of the ancient world, who studied in Alexandria as a youth. Eratosthenes, the historian and philosopher, was also there. He is perhaps most famous for measuring the size of the Earth by calculating the distance between two cities and observing the difference between the angle of the sun at solstice in them. He came within fifty miles of the correct figure. Hipparchus, who invented the concept of latitude and longitude, also resided in Alexandria. But perhaps the first and greatest of the scholars in residence was Euclid.

Euclid may have been one of the greatest mathematicians of his day, or he may have simply been an average mathematician who borrowed the ideas of others. We will never know for sure. Almost nothing at all is known about him other than a few sentences in a few ancient texts. So vague is the historical record on the man that some have debated whether he was a single person or a name used by several people collectively. His character, looks, manners, life's milestones, and even the dates of his birth and death are all completely unknown. His age can only be guessed at based on the fact that certain ancient references peg him as slightly younger than Plato and slightly older than Archimedes.

All that exists of Euclid personally are a few sentences that describe no personal attributes and only a few seemingly insignificant details of his life—some random exchanges and nothing more. One story, which Archimedes told, is of Euclid's conversation with King Ptolemy, who was speaking with Euclid one day about geometry, puzzling over it. He was complaining to Euclid about how hard it was to learn geometry and asked if there were not some easier way of tackling the subject. Euclid replied that no, there was not. "There is no royal road to geometry," Euclid said. The royal roads were famous in Persia for being straight and reserved solely for the king.

Another time, a boy who was learning mathematics from Euclid asked him one day, "What do I get by learning those things?" Euclid asked his slave to give the boy a few coins. "Since he must gain out of what he learns," Euclid said.

These two anecdotes illuminate the singular subject that defined Euclid's life—and the reason he is remembered. Around 300 BC, he wrote down everything known about geometry in his day in his book the *Elements*, a geometry textbook for the ages. It would

come to define geometry for more than a hundred generations—
more than two thousand years.

Elements is the most successful mathematics book of all time and
perhaps the most successful scientific book as well, both in terms of
its influence and its persistence through the centuries. It was so successful that it can still be found in print today. It was considered by
some to be more infallible than the Bible, and Euclid was its Jesus
figure. He is the only mathematician in history who has ever sum-
med up the entire field of mathematics in a single body of work.
Before him, many books of the same type on the same subject
were written. After him, none came for thousands of years. Euclid
eclipsed all previous mathematicians on the subject, and future generations would write little more than commentaries on his work.

Unlike many modern books, the *Elements* provides no soft introduction. There is no author's preface. No examples drawing upon
everyday experience are given to illustrate the underlying concepts.
Throughout, there is only cold, stepwise proof.

Philosophically the book was even more profound. Aristotle had
formalized the logical system, and after him, Euclid locked it into
place. Moreover, he conceptualized space and set it firmly in the
concrete, flat, three-dimensional world that we know. Space became
Euclidean because Euclid defined it as he saw it some twenty-three
hundred years ago.

The strange thing is, very little was new in the *Elements*. Euclid was
not the first Greek mathematician to write a book about geometry,
and he based much of his work on another book of the same name
by an earlier mathematician. He pulled together the work of his
predecessors, such as Eudoxus and Theaetetus. The tenth book of
the *Elements* is almost entirely the work of Theaetetus. None of the
other books called *Elements* survived through the ages.

Even if Euclid stole much, one cannot fault him as a mathematician. All indications are that he was clever and deserving of his fame.
That his portion of renown is perhaps greater than his ability as a
mathematician is irrelevant. Even if he was an accidental prophet of
mathematics, he was still a prophet. In any case, his book would
eclipse all others. In the two millennia that followed, it came to be
the canon of mathematics. Anyone who knew anything about the
subject was expected to know everything about Euclid's *Elements*. To

know geometry was to know the *Elements*. It was part of general mathematical curricula. It was also part of philosophy.

Even when more geometrical concepts were developed long after Euclid, they were not seen as something new but rather an appendix of sorts. After Euclid's death other books were written about geometry, but their authors wanted to be associated with the *Elements*. The fourteenth book of Euclid was written in the second century BC by a mathematician named Hypsicles. In the sixth century AD a Byzantine mathematician whose name is lost but who was believed to be the pupil of Isidoros of Miletus, the architect who reconstructed the St. Sophia church in Byzantium, wrote the so-called fifteenth book of Euclid, on solid geometry.

Writers would produce hundreds of commentaries on the *Elements* in the coming centuries, and the book was still so influential nearly two thousand years later that when Isaac Newton wrote his *Principia* in the early 1680s, he copied its style from the *Elements*.

Elements is actually thirteen separate books organized by subject matter and containing everything that was known about mathematics at that time in ancient Greece—the geometry of lines, circles, and solids, and a few books on numbers and ratios. Each book is filled with propositions, which are simply statements that tell you how to do things in geometry and related mathematical fields—how to draw different shapes, how to inscribe one shape inside another, how to find the geometrical relationships between various objects, lines, and angles, and so on. These propositions are all connected as part of a grand logical system that builds upon itself. Each time Euclid constructs a proposition, he follows it with a logical proof, often building upon one or more of the previous propositions.

As a starting point for all the proofs given in the first book at the very beginning of the *Elements*, Euclid wrote definitions and postulates at the beginning of each book. For instance, the first book starts with a set of twenty-three simple definitions that include various shapes like a point, a line, a plane, a circle, a diameter, obtuse and acute angles, and equilateral, isosceles, and right triangles. These definitions read almost like poetry. "A point is that which has no part/A line is a breadthless length/The extremities of lines are points."

From these definitions follow the first five postulates, which are basic rules established for the solely practical purpose of building the rest of the system logically. The first four postulates have a perplexing, even haunting, simplicity. The first three are concerned with the construction of straight lines and circles.

Postulate one says that it is possible "to draw a straight line from any point to any point," and is so simple it needs no explanation. Similarly, the second postulate, that one can "produce a finite straight line continuously in a straight line," simply means that you can extend a straight segment out farther. The third postulate may not be so obvious at first, but even a simple examination is enough to find it convincing. "To describe a circle with any center and distance" simply means that circles can be drawn, measured, or reproduced on the basis of a defined center and the circle's radius. Likewise, the fourth postulate, "that all right angles are equal to one another," is trivial. This is intuitively obvious, even.

The fifth postulate becomes harder to swallow. It says that if two lines are drawn that intersect a third in such a way that the sum of the inner angles on one side is less than two right angles, then the two lines inevitably must intersect each other on that side if extended far enough. Following this morsel are 465 theorems.

Nobody had a problem with the other postulates. But something about the fifth one made it different. There was no reason to accept it as true. Its converse is proven in the *Elements*, and if the converse is not self-evident and in need of proof, then why should the fifth postulate not have been the same? About two thousand years after Euclid penned it, Jean le Rond d'Alembert declared the fifth postulate the scandal of elementary geometry.

Sometimes referred to as the eleventh or the twelfth axiom, the fifth postulate is also an oddball because it is much longer and more involved than the rest. By the time of the Renaissance, it had become the thirteenth axiom. In it, Euclid basically says that two parallel lines will never meet, or rather that two lines that are not parallel *will* eventually meet in the direction where they are converging, or moving toward each other.

The word *parallel* derives from the Greek word meaning "running alongside." According to the fifth postulate, if two lines are not running alongside, they are running together. Which side converges can be ascertained if another line is drawn across the two. "If two lines are drawn which intersect a third in such a way that the sum of the inner angles on one side is less than two right angles," Euclid wrote, "then the two lines inevitably must intersect each other on that side if extended far enough."

This is perfectly logical to assume, since if you draw two lines in a plane, they will either be parallel or not. Logic says that if two lines are converging (angled toward each other), if you draw them out far enough they will eventually meet. It's easy to imagine. All you need is a space big enough. Even if you have to draw the lines out past Jupiter, converging lines have to meet eventually.

In modern science, nobody is expected to accept something on faith. A hypothesis remains a hypothesis until it is tied to observational data and sufficiently explains that data enough to become a general theory of how that little part of the universe works. But how can one prove the fifth postulate without actually drawing the lines? It's not hard to convince yourself that the postulate is true using common sense. You could consider two lines drawn to infinity, but can you logically, mathematically prove that they do or do not meet? It's easy to imagine lines stretching off into infinity, but it's impossible to actually see them. Other than in the imagination, there is no easy way to prove the postulate. This did not stop people from trying, of course. Mathematicians before Euclid had tried to do so. Euclid himself tried but couldn't do it. Some of the most famous mathematicians after him tried to settle the question. They did not doubt for a second that it could be proven, but they failed to do so. Everyone everywhere failed.

Why did the fifth postulate need to be proven? Why not just accept it without proof? Can't one simply assume and accept that converging lines will meet eventually?

Unfortunately, that assumption does not work for the fifth postulate, and the Greek mathematicians were too smart to accept it. The Greeks were already familiar with a curve known as the hyperbola, for instance. One type of hyperbolic curve is typically represented by the equation $y = 1/x$. When x and y are graphed in a modern coordinate system, the resulting curve looks like a boomerang that stretches out to infinity. The arms of the curve constantly approach the horizontal and vertical axes, but they never touch. As x gets larger, y approaches zero. Conversely, as x gets smaller and smaller, the value of y approaches infinity. The value of y never quite reaches zero or infinity in either case. The Greeks knew that. The fact was, the fifth postulate needed to be proven.

The problem is that the postulates are not there by accident. They are starting points, the necessary assumptions upon which the rest of the book is built. The postulates are the foundation blocks. They are the lowest stones in the wall that bear the weight of all. They need to be statements that are so obviously true and necessary as to be acceptable without proof.

The third postulate, for instance, says that one can to construct a circle by defining its center and specifying its radius. If you cannot accept this, how can you go on to construct numerous circles and explore their properties? Likewise, the fifth postulate is relevant because it is used to prove other results. Without it, much of the first book of *Elements* could not be written. The fifth postulate implies that there is only one parallel to a line through a point not on that line. It is used to prove that two parallel lines are everywhere equidistant. It is used to prove that the three angles of any triangle add up to 180 degrees. Proving this, however, requires the fifth postulate to be true. And because it had to be true, it needed to be proven. Without proving it, how could one know for sure?

Almost from the moment Euclid finished his *Elements*, its readers began to see a problem with the fifth postulate. Many mathematicians for the next two millennia, out of a desire to simplify the argument and solidify the logic, tried and failed to prove it. Euclid himself may have abandoned his own attempts and resigned to write the postulate in such a way that it didn't need to be proven. It is impossible to know what Euclid felt about his own book, but the work was not held in as high esteem during his lifetime as it was after his death. He became wildly famous after he died, and this ratcheted up the pressure to prove the fifth postulate.

Every generation of mathematicians after Euclid believed in him and subscribed to the same basic method of systematic proof. Medieval scholars used this method of proof to demonstrate the existence of God. Some even claim that the Declaration of Independence owes its eternal form to the Greek method of proof exemplified by Euclid. "We hold these truths to be self-evident," it begins. And so it was for Euclid. His postulates were meant to be self-evident truths.

Perhaps it was better for Euclid to leave the fifth postulate as a mysterious patch of wilderness—a dark area from which old mathematicians could warn away their younger protégés. Somehow this was not good enough. Geometry, after all, was the science of the real. Plato called it the "knowledge of that which always is."

What was supposed to be forever lasted for a little more than two thousand years. As Gauss and Farkas Bolyai's son János would soon discover, there was no reason that the fifth postulate *had* to be true, and there was no reason that geometry had to be that which always is.

4

Those False and Would-Be Proofs

It is ignorance alone that could lead anyone to try to prove [an] axiom.
—Aristotle

They were the loneliest mathematicians in history, struggling to prove something that could not be proven. Some gave up in agony and failure. Some toiled throughout their lives. Many died aware they had not solved the mystery. Others happened upon a solution—or so they thought. After years of efforts, they may have satisfied themselves that they had found it, only to be proven wrong by the people who followed in their wake.

Even so, the mystery of the fifth postulate was seen as unfinished business rather than a flaw in Euclid's great work. Until the early part of the nineteenth century, nobody had considered that the postulate might be false. In fact, the opposite occurred—nothing seemed truer than the fact that two parallel lines would never cross. When England's Robert Recorde introduced the equals sign (=) in 1557, he did so intending to represent two parallel lines. "Because no two things can be more equal," he wrote.

Euclid himself may have struggled for countless hours over proving the fifth postulate or on whether he even needed to make it a postulate at all. We will never know. What we do know is that Euclid did not include this postulate among the originals. Modern scholars discovered that the fifth postulate did not appear at the beginning of Euclid's *Elements* at all, but later on—and only when it became absolutely necessary to prove some of the propositions in the book, such as the angles in a triangle add up to 180 degrees.

So after Euclid, the question slowly arose as to whether the fifth postulate could be proven by taking into account the other postulates. Mathematicians tried to prove it many different ways. Some attempted to reformulate or redefine what parallel meant. Others sought to restate the postulate somehow. Some mathematicians thought about the nature of the straight line and tried to use their reasoning there to prove it. Still others tried to prove the fifth postulate by building upon the definitions and the first four postulates of Euclid's geometry as the basis for the proof.

It is hard to say exactly how many people wandered down this lonely road. There are dozens whom we can point to today—mathematicians whose fame caused their ill-fated work on the problem to be recorded in history. But the attempts we remember are only the cream on top of a bucket of forgotten failures, such as that of Claudius Ptolemy (the mathematician not the king), who studied at the academy in Alexandria in the first half of the second century AD. He was the second most influential thinker of the ancient world behind Aristotle, and perhaps the most famous mathematician to work on the fifth postulate. His proof was really only an elaborate restatement.

There were many more besides, including a few who saw the problem clearly and made valiant attempts to solve it. Around the beginning of the first century BC, the philosopher Posidonius thought he could prove the fifth postulate based on the definition of parallel lines. Specifically, he wanted to change the definition of a parallel line to include the concept of equidistance. By stating that parallels are everywhere equidistant (and therefore never meet), he thought he could render the problem unnecessary. Since they are equidistant in one place, they are equidistant in all places. This was a clever solution, but it missed one critical point: it did not refute the possibility that there might be other lines that were not parallel and also did not meet. The fifth postulate says that lines that *are not* parallel will cross. Changing the problem by retelling it in another form (lines that are parallel *will not* cross) still does not prove the postulate, though, because it requires an assumption equivalent to the postulate.

Posidonius had a pupil named Geminus of Rhodes who studied the fifth postulate around 70 BC and recognized that it still needed to be proven. Geminus may not have been the first to consider a more elaborate proof of the postulate, but he is the first person whose dissatisfaction with it is recorded in history. Though Geminus's work is lost today, later Roman scholars quoted him and his attempts to prove the fifth postulate, the earliest work on the problem for which the details are known. What comes through most from this lost work is the sense that Geminus had profound contempt for the fifth postulate. He pointed out that it is plausible but that there is no reason to believe it is necessary. He knew it had to be proven, but he could not do it.

During the centuries of Roman dominance, work on the fifth postulate—and mathematics in general—slowed. The Romans sacked Syracuse in 212 BC, and in the process they killed Archimedes. He built machines to help fortify the city and keep the Romans at bay for a long time. When the city fell, the Roman generals wished to capture him and put his genius to work for them. But he was killed, probably struck down by the sword of a Roman soldier who had no idea who he was.

The death of Archimedes at the hands of an oblivious soldier is the perfect metaphor for how the Romans treated mathematics in general. They had little interest in knowledge for its own sake. They were mainly interested in mathematics insofar as it could be used to further one objective—the advance of the Roman state. They did not support its development independent of this. In fact, the Romans regarded pure mathematics with a mixture of suspicion and contempt.

Because of this, they contributed little to the advancement of the discipline. Some even say that in hundreds of years, the Romans contributed nothing of value, not a single noteworthy work. In those lands that Rome conquered where mathematics had flourished for half a millennium, the subject began to dwindle.

If the Romans excelled at anything, it was at appropriating the ideas of others. They learned architecture and fashion from the Etruscans. From the Phoenicians they borrowed glassblowing. The Egyptians taught them to work with dyes. From the Greeks they learned weaving, toolmaking, mathematics, a few key cultural ideas, and philosophy.

The Greeks achieved more than that. Rome established supremacy over the Greek world, but Greece reigned over the Roman mind. The Romans had modeled themselves on the Greeks from the very beginning. Soon after the Roman Republic was established in 510 BC, the rulers sent a mission to Athens to study Greek law.

In the end, though, the Romans were exceedingly practical people, and their main objective was to consolidate and advance their own society and power. Mathematics for them was essentially limited to accounting and surveying, and so under Roman rule, mathematics more or less reverted to the practical art that it had once been under the Egyptians—simple rules that could be applied to immediate uses. Geography, for instance, became a tool for the generals to guide their armies. Centuries after rising to dominance over the Greeks, the Romans were still far behind them in certain ways.

Some great mathematicians still existed. One of the greatest was Hero of Alexandria. His work was based mainly on that of Euclid and Archimedes, and he made few important discoveries of his own. His emphasis seems to have been in surveying for engineering purposes and in mechanical devices used in warfare. Another two noteworthy mathematicians, toward the end of the fourth century AD, were Theon of Alexandria and his daughter Hypatia, the last of the great mathematicians at Alexandria. Together they wrote a version of the *Elements* in Greek with numerous corrections and Theon's own commentaries.

Theon, living seven hundred years after Euclid, did not like the version that existed in those days, so he rewrote it, clarifying the language. Somewhere in working out the translation or the commentaries, Theon and his daughter decided to take a much more vigorous approach to adapting the book to the fourth-century student, and they altered parts to make them stylistically consistent and in some places more simple. They added alternative proofs and minor theorems and also included elementary explanations of the proofs. Theirs became the standard Greek edition of the *Elements*. So synonymous did Theon become with Euclid's work that for centuries afterward he was regarded as almost equal in importance to Euclid in writing the text. People believed that Euclid had come up with the propositions in his book and Theon proved them. Some even thought that Theon was the real author of the *Elements*. His daughter Hypatia, despite her brilliance, suffered a terrible fate. She was murdered in 415 AD, torn apart by an enraged mob.

A history of attempts to prove the fifth postulate was written by a mathematician named Proclus in the fifth century. This was perhaps his most interesting work. One of the most important thinkers ever to stumble upon the mystery, Proclus wrote extensive commentaries on Euclid and other great thinkers from the golden age of Hellenistic culture. He may have written comments on all of the books of the *Elements,* though they do not survive today. The only one that did was his commentary on the first book, in which he discusses the fifth postulate.

He shot down the notion advanced by some that the fifth postulate was self-evident. Since the hyperbolic curve approached a line but never reached it either, he asked, "May not this, then, be possible for straight lines, as for those other lines?" In fact, Proclus refused to even consider the fifth as one of the postulates. It "ought to be struck from the postulates altogether," he said. He also called attention to the circular logic of earlier attempts to prove it. He discussed Ptolemy's attempt, and showed that Ptolemy was wrong. Finally, Proclus discussed how it wasn't even necessary to assume that the fifth postulate is true. Saying so didn't stop him from trying to prove it anyway, though. In one attempt he assumed that if a third line were to be drawn in the same plane and cut one line, then it could cut the other. But this was just yet another retelling of the postulate.

Proclus attempted another proof of the fifth postulate wherein he examined how there is a finite distance between two parallel lines. Given this finite distance, a line that crosses one line must also cross the other, he claimed. The problem with Proclus's second proof is that it also begged the question. His initial assumption was that the distance between any two parallel lines must be finite, but there is no reason that this must be true.

In the next few centuries, geometry slowly ceased to be taught in Europe. Schools closed and teachers died, and interest in the subject died along with them. The emperor Justinian closed the Athens Academy in 529 AD, one of the first acts in his long reign. As easy as it is to focus on that, there was no single event that robbed Europe of geometry. Rather, it was a slow, creeping death that unfolded over time. The subject might have been lost forever if not for intense interest by scholars in the Middle East. Geometry was exported there, and it was studied and passed on by legendary Arabic scholars. The time of Muslim mathematics had arrived.

The first Islamic scholar to translate Greek works into Arabic was a man named Khalid ibn Yazid. Beginning around the start of the eighth century, he encouraged Greek philosophers in Egypt to translate the Greek works. By the middle of the century, all the important work in mathematics was being done by Islamic scholars. This would remain the case for the next several hundred years.

Toward the end of the eighth century, a scholar named Muhammed al-Fazari translated works from Sanskrit and introduced Hindu numerals into the Muslim world. This event had profound implications for the history of mathematics because the Hindu numerals (later called Arabic numerals), being much easier to work with than Roman numerals, would change the nature of mathematics. An equally profound change was the introduction of a vast body of Greek literature into Arabic and Persian culture.

When the seventh Abbasid caliph, Abdallah al-Ma'mon, ascended to the throne in 813, he changed the learned Arabic world. He was a strange contradiction—learned and scholarly and yet strangely and violently intolerant of dissent. He founded a school and encouraged its greatest teachers to translate the works of the pagan Greek writers, and yet he cruelly persecuted fellow Muslims who disagreed with his views of the Koran. He surrounded himself with genius. Jew and Christian alike were welcome in his court.

Al-Ma'mon sent a mission to the Byzantine emperor Leon to obtain Greek manuscripts, and in doing so he cemented a second, earlier renaissance. Scholarship exploded in the Middle East, fueled in part by the availability of numerous ancient texts from Byzantium and elsewhere. In the ninth century and afterward, many of the classical works of the ancient world were translated from Greek into Arabic. At the center of this renaissance was Baghdad, which enjoyed the perfect combination of smart, interested scholars and wise, interested patrons.

The three Ibn Musa brothers donated the greater part of their wealth to acquiring and translating these texts. Under their patronage, Ptolemy's famous book the *Almagest* was translated into Arabic. *Almagest* is an Arabic word that means "very great." Euclid's *Elements* was also translated. It was of special interest to two of the brothers, who became students of geometry. The first translation of the *Elements* into Arabic had been made around 786 AD by a man named al-Hajjaj ibn Yusuf. Around the same time another mathematician in Baghdad named al-Abbas wrote commentaries on the *Elements*.

With the added interest of the Ibn Musa brothers, work on the *Elements* flourished.

Something that helped this flowering was a project the caliph al-Ma'mon undertook at Baghdad. He opened a school there called the House of Wisdom, and out of this academy came Arabic translations of many of the important works of the Greek and Roman times. This was the most ambitious educational project in learning since Ptolemy Soter founded the school at Alexandria more than a thousand years before.

From this school arose some of the greatest Persian and Arabic scholars of all time. One of these was Abu Yusuf al-Kindi, who wrote more than 250 works on everything from math to music to astrology and Aristotle. Almost none of it survives, but he had a massive influence on Persian, Arabic, and European thinkers who followed him, since some of his works were later translated into Latin. Another scholar named Hunain ibn Ishaq had lived in Alexandria and learned Greek, making him an excellent translator.

In the middle of the ninth century, Hunain was hired by the three Ibn Musa brothers to begin translating books. He went to Byzantium with a group in search of manuscripts to be translated, and he later made another translation. Hunain took great pains to find good manuscripts, comparing them to versions already translated into other languages.

An Arabic mathematician and linguist named Thabit ibn Qurra was another great translator. He even founded his own school of like-minded translators who strove to bring the works of the ancient world into his modern Arabic one. He also would revise and improve the translations that others made, including of the *Elements*. Thabit was a talented linguist and mathematician, which made him uniquely suited to this work. All later versions of the *Elements* derived from his work.

The translators were joined by a great succession of thinkers and commentators. By the mid-ninth century the mathematician al-Khwarizmi synthesized Greek mathematics with Hindu numerals. Modern recognizable mathematics was born. Al-Khwarizmi wrote a book called the *Hisab al-jabr wal-mugabala*, the short title of a work that translates as *The Compendious Book on Calculation by Completion and Balancing*, and from the *al-jabr* part of the title the word "algebra" was coined. He did much original work and was the most influential mathematician of the Middle Ages. His work was later discovered and translated into Latin by Europeans.

The mathematician al-Mahans wrote more commentaries on Euclid toward the end of the ninth century. So did al-Nairizi a generation later. Al-Nairizi's commentaries were eventually translated into Latin. Abu Uthman translated Euclid's book ten and commentaries on it in the beginning of the tenth century. At the end of tenth century more translations of Euclid were made by Muslim scholars. Al-Khazin, al-Saghani, and Nazif ibn Yumn all wrote works on geometry, some direct translations of Euclid's book, others commentaries and extensions of it. One of the last translators and commentators was Abu-al-Wafa, who wrote commentaries on Euclid and helped develop trigonometry.

By the end of the tenth century, the great work on mathematics was spreading. At the very end of the century the French mathematician Gerbert rose to become Pope Sylvester II. He had spent a few years in Barcelona earlier in his life, and it may have been there where he came into contact with Arabic numerals.

In Spain, the scholarly tenth-century ruler al-Hakam was a patron of science. His personal library was said to contain 400,000 volumes, and he sent agents all over the Muslim world to find and translate texts. Books written in Asia were known to al-Hakam in Spain before even Asian scholars knew about them. Not surprisingly, mathematics flourished in Spain during the eleventh century.

One of Gerbert's students, named Beruelius, and others began to translate and comment further on Arabic works. Slowly it became acceptable to plumb the Muslim world for learned works. There was a lot to learn at that point.

In the early part of the eleventh century, the most famous scientist in the Islamic world, Ibn Sina, or Avicenna as he was known in Latin, wrote a translation of Euclid along with his voluminous works on medicine, philosophy, and physics. Avicenna was so prolific and learned that some considered him unnatural—a magician.

Egypt, too, had another mathematical renaissance. The greatest astronomer of Arabia—and perhaps the greatest Muslim astronomer of all time—was Ibn Yunus. He prepared careful tables of astronomical observations in the late tenth and early eleventh centuries thanks to a new observatory built for him in Cairo. There a great school grew up over the next century called the Hall of Wisdom—second only to the House of Wisdom.

By the dawn of the twelfth century, there were still Muslim commentators on Euclid's book. One of the last of these was Muhammed ibn abd al-Baqi, whose study of book ten of the *Elements* would soon be translated for readers in Europe.

In the thirteenth century, some Muslim scholars, such as Nassiruddin at-Tusi and the Persian mathematician and poet Omar Khayyam commented on the fifth postulate and tried to solve the mystery. Omar Khayyam was the greatest mathematician of his day, and he advanced algebra considerably and studied Euclid thoroughly. He accomplished nothing by way of proving the fifth postulate. Still, his work was later printed in Rome and studied by others who were interested in proving it.

Nassiruddin was an interesting character who led a rich life. Born in the city of Khurasan in the northeast corner of Persia around 1200, he was one of the most influential writers on philosophy, mathematics, and a number of other subjects in the thirteenth century. He wrote dozens of major works in a large number of areas, including an edition of Euclid with extensive commentaries.

His life took some strange twists. He was kidnapped by the local governor and handed over to the grandmaster of the powerful Assassins clan and brought to their mountain fortress at Alamut. The grandmaster assassin had arranged the kidnapping because he wanted to adorn his court with the greatest scholar in his land. Nassiruddin was an honored guest and a valuable prisoner at the same time.

He was also a trusted adviser, and when the Mongol hordes advanced upon Persia, the grandmaster assassin turned to him for advice. Surrender, said Nassiruddin. This was the best course of action. The Mongols were famous for sacking towns that resisted them. The greater the resistance, the more the raping and pillaging. And so far, nobody had been able to stand up to the Mongols. This made sense in a way, and so convinced, the grandmaster of the assassins surrendered, the Mongols put him to death instantly, and they took Nassiruddin as one of their prizes. Nassiruddin became a sort of soothsayer to the Mongols. He was an expert at astrology and at interpreting patterns of lines in the sand. He was also an authority on ethics and the author of one of the best-known works on the subject.

But with the Mongols, Nassiruddin witnessed the rape of Baghdad. When the last Abbasid caliph was brought before the Mongol leader Hulagu, Nassiruddin was there. He convinced Hulagu that there would be no divine punishment for putting the caliph to death. The legend is that the Mongols were superstitious about spilling the caliph's royal blood, so they rolled him up in a carpet and let horses trample him.

Of note is Nassiruddin's work on the fifth postulate. He had worked out an incorrect proof based on the sum of the three angles of a triangle and tried to reduce the fifth postulate to a simple, equivalent argument. He showed that the fifth postulate could be proven if the equivalent could be proven: that the sum of the interior angles of a triangle is equal to 180 degrees. Proving this fact would be enough to verify the fifth postulate, Nassiruddin sought to demonstrate. Unfortunately, he was not able to do so, and ultimately he failed.

Nassiruddin gave another equivalent to the fifth postulate. He said that if all the points on a curve were equidistant from the points on a straight line, then that curve would be a straight line. He was also important in the history of geometry because his work was later rediscovered in Europe and extended. By the time Nassiruddin died, Europe was already awash in Arabic scholarship, newly translated into Latin and other languages. It is hard to say when the translations began appearing, because in those days before the printing press, mass production was impossible, and books were copied by hand and distributed piecemeal. Perhaps the diffusion of Arabic works had been going on for centuries.

The Europeans had certainly enjoyed other fruits of Muslim lands. Chief among these was the abacus. Initially, the device was used with Roman numerals, but this, too, would soon change due to Muslim influence. Its use grew steadily over the eleventh century, and it really took off during the mid-twelfth century. By then, the Arabic versions of Euclid's *Elements* had been transported to North Africa and Spain with Muslim scholars, and from there it leaked back into Europe.

In 1120, an Englishman named Adelard disguised himself as a Muslim scholar and traveled in Arab lands. He was from Bath in England and had gone to school in France and taught in the French city of Lyon for a time. From there he spent several years traveling. He visited Italy and Sicily, which had been conquered by forces from North Africa and were under the control of Arabs for much of the previous century. This led him to continue exploring the Muslim world.

He traveled to Turkey and Syria, where he came into contact with all of the ancient Greek works of mathematics and astronomy as they existed in Arabic translations. And at some point he landed in Spain, where he made one of the greatest contributions to European enlightenment. There Adelard came into contact with the ideas and teachings that were popular among Arabic culture at the time, including mathematics. He found a copy of the *Elements* in Arabic, which he translated into Latin. He translated two different versions of the book plus a commentary, which became known as his own version.

By 1126, Adelard was back in the West. He might have been remembered for his statement that he preferred reason to authority, or for his other books on arithmetic, music, and astronomy, but it was his translations of ancient mathematical texts that brought him fame. They remained in use for centuries in Europe. He found the work of al-Khwarizmi and translated it into Latin. This work was subsequently copied over and over, along with many other texts he translated. Adelard was not alone. Over the next twenty years, a large group of scholars in northern Spain worked to translate more texts of the Greeks and Arabs into Latin.

Europe slowly rediscovered Euclid in the thirteenth century. It was as if the continent had been in a long slumber. When Greek mathematics and science were returned to the West, they were enriched by the addition of all of the Arabic commentaries that had been written through the years, including those on Euclid's *Elements*.

In England, Roger Bacon started reading geometry from Adelard's books and embraced it. Universities in Europe, which had been founded at the beginning of the twelfth century as meeting places where students could gather and learn, were now thriving. One of the great subjects studied was mathematics. By the end of the thirteenth century, students at Oxford were endeavoring to learn Euclid's definitions and theorems.

This was an exciting time for mathematics for another reason. At the dawn of the thirteenth century, Leonardo of Pisa, the most influential mathematician you may have never heard of, brought a great revolution to mathematics in 1202 by introducing Hindu-Arabic numerals into the West—one of the great innovations in history but also one of the simplest. With Arabic numerals came a greater facility to do mathematics, and thus a greatly enhanced interest in the subject.

In the thirteenth century, Campanus of Novara published a beautifully adorned Latin edition of the *Elements* that later became the standard printed version in Europe for centuries. He is a mysterious figure in history, who was supposed to have been the chaplain of Pope Urban IV from 1261 to 1264. Campanus revised Adelard's translation by drawing upon other sources that were then available. Many of the early editions of Euclid's *Elements* carried a commentary by Campanus of Novara.

Work on geometry continued. In the early fourteenth century, a scholar named Levi ben Gerson wrote an important commentary on Euclid's *Elements*. In 1482, ten years before Columbus sailed to the New World, the first printed copy was made in Venice. For more than fifteen hundred years, books had been copied by hand one at a time. Now they could be mass-produced. In 1533, Proclus's commentaries were published and began to be circulated among the European universities. Robert Recorde in England began writing mathematical books in English and in 1551 produced his *Pathwaie to Knowledge*, a simplified abridgement of the *Elements*. Sir Henry Billingsley, who later became the mayor of London, made the first translation of Euclid into English in 1570. But through these centuries, no Europeans took on the challenge of trying to solve the fifth postulate.

That all changed in the beginning of the seventeenth century. The first attempt came in 1626 by an Italian mathematician named Pietro Antonio Cataldi. His reasoning was based on equidistant lines, and he postulated that lines that were not everywhere equidistant would converge in one direction and diverge in another.

He came no closer to solving it, though. Cataldi simply reformulated the fifth postulate into a new statement that was no closer to proof than what Euclid had written two thousand years before. The problem with this, as before, was that there was no more reason to accept Cataldi's postulate without proof than there was to accept Euclid's. Still, the seed was planted.

The revolution in astronomy in the sixteenth and seventeenth centuries made Euclidean geometry more important than ever. For more than a thousand years astronomy had suffered under the worldview established by Claudius Ptolemy, who put Earth at the center of the universe and had the heavens, the sun, the moon, and the planets revolving around us. The Ptolemaic system was an elaborate and unwieldy one, because the planets did not move in simple paths across the sky that would suggest a simple orbit. This required the theory to be modified by introducing a complicated correction known as epicycles. As Sir Arthur Eddington once said, "The cosmogonist had to fill the skies with spheres revolving upon spheres to bear the planets in their appointed orbits; and wheels were added to wheels until the music of the spheres seemed well-nigh drowned in a discord of whirling machinery."

Along came Nicolaus Copernicus, who turned the solar system on its head by revising the Ptolemaic system and designing a solar system in which the planets revolved around the sun. The beauty of the Copernican system was that it greatly simplified planetary motion. This was advanced even further by Johannes Kepler, who derived the basic laws that described orbital motion and helped to free astronomy from speculation and personality.

Kepler once said that his aim was to show that the true nature of the heavens did not resemble an animal but rather a clock. "Astronomers should not be free to feign anything whatever without sufficient reason," he wrote. "You ought to be able to give probable reasons for the hypotheses you propose as the true cause of appearances."

In the seventeenth century, things really began to change. Mathematics was finally beginning to move ahead into new territory and to surpass the knowledge of the ancients. This was one of the most exciting times in the history of science. The whole field of mathematics was advancing as never before. Sir Henry Savile was lecturing on geometry at Oxford. Similar lectures were being delivered at Cambridge. Astronomers after Johannes Kepler sought to explain the motion of bodies in the heavens in terms of mechanical motion through physical space.

Euclid's geometry was what defined that space. The desire to prove the fifth postulate became even more pronounced after René Descartes revolutionized all of the mathematical sciences by introducing his Cartesian coordinate system in his 1637 book *Geometry*, an innovation that has been likened to a violent act for how radically it affected things. Fourier once said of Descartes's work, "There cannot be a language more universal and more simple . . . and better adapted to express the invariable relations of nature." Descartes also formalized the notion that space was infinite, without limits in its extension. Geometry was the basis of physical space and it was Euclidean in nature.

All of the great mathematicians of the seventeenth century— Johannes Kepler, Galileo Galilei, John Napier, Pierre de Fermat, Descartes, Blaise Pascal, John Wallis, Christian Huygens, Isaac Newton, Gottfried Leibniz, and others—lent credence to Euclid's version of geometry either directly or indirectly. Proving the fifth postulate became more important than ever during the seventeenth century—especially after Isaac Newton applied geometry to the physical nature of objects in space. Euclidean space was physical space. Geometry was to physics as anatomy was to medicine.

Still, there was one fundamental flaw in this physiology, and one of Newton's contemporaries thought he could solve it.

5

A Codebreaker's Fix

It hath been my lot to live in a time, wherein have been many and great changes and alterations.

—John Wallis (1616–1703)

The Oxford professor John Wallis was not the most likely person to solve the greatest mathematical mystery of his day. He was not even really a mathematician.

Wallis was a divinity scholar, a well-heeled theologian, and that's all he might have been had he lived at another time. But he was a product of his times. In the mid-seventeenth century, England was a country torn apart by internal strife. The times thrust young Wallis into greatness. His interest turned toward mathematics, and this would bring him as close as anyone had ever come to proving Euclid's fifth postulate, an unsolved problem that was as old as Plato.

Barely out of university when the English Civil War broke out in 1642, Wallis gravitated to the revolution. Chaotic as those times were, they rewarded people like Wallis. He offered specialized skills to the cause in his off hours.

Wallis discovered almost by accident that he had the ability to decipher code. He was, in fact, extraordinarily good at it. He devoted energy to the Protestant cause by interpreting intercepted enemy communiqués. This was not always an easy task, but it needed doing if any military or political intelligence was to be gained. Wallis was more than equal to the task, as he proved on numerous occasions.

His deciphering career started one night, just after the Civil War began, when he was in London having dinner at the house of Lady Mary Vere, a wealthy widow. Wallis served as her private chaplain—something that well-educated ministers often did for the well-off in those days. That particular night in 1642, another chaplain was also present. He served as minister to a volunteer general in the Puritan army.

The rebellion had only just started, and warfare with King Charles I and his Royalist army was both open and open-ended. Things didn't necessarily look good for the rebels. The king's forces, after all, were professional soldiers and among the finest in Europe—seasoned, trained, disciplined, and paid. Compared to them, the rebel forces were somewhat of a rabble. If the rebel cause was to survive, it needed to exercise any advantage it could. And of course, as in any war, one of the greatest potential advantages is intelligence.

The general's chaplain mentioned to Wallis that a letter had recently been intercepted en route to Royalist forces encamped away from London. But the letter was written in code, and nothing could be made of it. Perhaps its contents could assist in turning the tide—however slightly—by offering intelligence that could be of aid to the rebels.

Wallis, a young man full of energy and passion for the Puritan cause, boasted that he could accomplish the task. At the very least, he volunteered, he would give it a try. So someone ran off to fetch the letter. And when dinner conversation ended and the guests parted ways, Wallis retired for the evening. He sat down, and in about two hours' time, as he later wrote, "I had deciphered it."

He sent the decoded message along the next morning. Thus began his career as a codebreaker—a career that soon enjoyed another triumph. A few months later, another communiqué was discovered. This one seemed even more important. It was sent by a former member of the king's inner circle—a man named Sir Francis Windebank, who was holed up in France. Windebank was attempting to get a message to his son in England using a highly complicated numeric code. Could Wallis decipher it?

Not immediately. But after a few months of periodically working on it, he cracked the cipher and disclosed its message. Wallis's reputation as a codebreaker was confirmed. For the next several years, he regularly decoded secret letters for his parliamentary allies.

On numerous other occasions throughout his career, Wallis was called upon to repeat his wizardry on other communiqués that had been intercepted. So successful was he at this that he once boasted in a letter that he had decoded enough of King Louis

XIV of France's military designs for conquering the German states to make a significant contribution to their continued existence. By the end of the century, he was known throughout Europe for his skills in cryptography. In 1699, his acquaintance the philosopher and mathematician Gottfried Leibniz referred to him as the most famous codebreaker in Europe.

Wallis personally profited from this work as well. Steeped in the political intrigues of his day, he would become thrust into the insular world of elite academia thanks to his deciphering abilities.

In recognition of Wallis's efforts toward Parliament's victory in the English Civil War, Oliver Cromwell grandly awarded him the Savilian chair of geometry at Oxford University in 1649. Sir Henry Savile had been a tutor of Queen Elizabeth, and the chair was richly endowed with the proceeds of the benefactor's lands in Essex, Kent, and Northampton. It was one of the most prestigious professorships in those days. The former holder of this post, a man named Peter Turner, had been unceremoniously ejected from the chair because he was a Royalist and had enthusiastically supported the losing side in the war—to his ultimate misery.

When the English Civil War broke out, Turner left the university to fight on the king's behalf. But the war made a cruel waste of his life. He was captured after one of the very first battles, thrown in prison, and stripped of his possessions. His academic career ended slightly less abruptly than his stint as a soldier, but it was just as doomed. Turner was ordered removed from his professorship at the end of 1648. Wallis moved into the position the following year.

Many must have seen Wallis as a poor choice for the chair. He was a codebreaker, sure, but he was no mathematician—certainly not when he was first appointed. Though he had been well educated privately and could boast degrees from Cambridge, his training was as a clergyman, and he had no formal schooling in the field in which he was now to be an expert: geometry. Nevertheless, Wallis served as a valuable and successful member of the university for half a century. If he started out as the Savilian professor with some doubters, he served so effectively that when he died he left only admirers behind.

He was an ambitious teacher and scholar and made interesting discoveries in many fields. He even invented a way of teaching deaf-mutes. He wrote numerous theological texts, and might in some ways had been the model for the modern would-be religious scholar—brilliant, prolific, and profoundly devout. The subject that defined his career, though, was the one to which he was appointed—geometry and mathematics. He was on the cutting edge of English mathematics. Wallis wrote a number of important mathematical texts in his day, including the highly influential *Arithmetica Infinitorum*—a book in which modern mathematics was struggling to emerge, as one modern writer put it. It influenced both Newton and Leibniz in their development of calculus. A younger contemporary of Wallis's named David Gregory saluted the *Arithmetica Infinitorum* as the foundation upon which all later discoveries in the field were built. By the time he died, Wallis was a living legend and perhaps would have been the greatest mathematician in his era had his life not overlapped with Isaac Newton's.

But perhaps the strongest testimony to Wallis's skills was that he managed to avoid the fickle fate of some after the Commonwealth government crumbled a decade or so later. After England's lord protector Oliver Cromwell died in 1658, his son Richard ruled for a few months but could not keep power, and he soon stepped down. Shortly after that, the English monarchy was restored and the king returned to the throne. King Charles II was no stranger to bitterness over the deposition and execution of his father, and he had Oliver Cromwell dug up and his head put on a pole and displayed outside Westminster Abbey for years.

These were worrying times for former supporters of the Commonwealth government like Wallis. Many were threatened by the return of the king. Despite the reversal of fortunes, though, Wallis did not lose his plum professorship—instead, he was reconfirmed to his post. He survived any would-be purges, and he even became the chaplain to Charles II. He only vacated his posts with his death in 1703.

As Wallis's skills had served the forces of Parliament during the Civil War, so would they serve the king after his restoration. And in 1663, just a few years after King Charles II returned to the throne, Wallis set his sights on breaking another, far greater code.

The year 1663 was a brief respite from the chaos of the previous two decades—two of the most turbulent in England's history. It was a relatively quiet time, and this gave Wallis time to think about triangles—key, he thought, to breaking the greatest code in the history of mathematics, a problem with no solution that had lingered for nearly two thousand years.

Wallis was enamored of the ancient thinkers. Their shadow covered anything that he ever laid his hand upon. An extreme expression of this was his vehement opposition to England's adoption of the new Gregorian calendar. He fought vigorously against replacing the older and less accurate Julian calendar, raising a number of historical, political, scientific, and religious objections against it (associated with Catholicism as it was)—successfully, too, it seems, for England did not adopt the Gregorian calendar until long after his death and more than a century after most of the rest of Europe.

In mathematics, too, Wallis exalted the old ways. He once wrote that he sought to "examine things to the bottom; and reduce effects to their first principles and original causes. Thereby the better to understand the true ground of what hath been delivered to us from the Ancients, and to make further improvements of it."

That was why he must have been excited in 1663 to be on the verge of what he must have felt was a solution to the oldest mystery in his field of geometry—proof of the fifth postulate. Wallis was influenced by Nassiruddin, and he wrote a commentary on that scholar's attempt to prove the fifth postulate. Nassiruddin had said that the postulate could be proven if you could prove that the angles of triangles always sum to 180 degrees.

Wallis's attempt to prove the fifth postulate relied upon the idea of similar triangles. If you take two lines that cross each other and then you cut them by a third line, as Euclid did in his postulate, you form a triangle. It didn't matter how large the triangle was—it might be one in which two of the sides measure out past the planet Jupiter—it would be a triangle nonetheless.

And for any given triangle of however near-infinite magnitude, Wallis speculated, there existed a similar triangle of whatever arbitrary magnitude one chooses. Triangles large and small could be similar to one another, Wallis knew, and the significance of this was that a triangle of massive scale could be represented by a much smaller triangle—one with proportionality such that it could be handled more easily. Thus for an infinitely large triangle, you could construct a smaller triangle, and prove the fifth postulate. Wallis conjectured that similar figures existed and that the existence of one triangle was enough to demonstrate the possibility of the other.

This was as close as anyone had come to proving the fifth postulate. It was not a proof, however, and as great a mathematician as Wallis was, he had failed. What he didn't realize was that he had merely restated the fifth postulate in another form. The problem with his proof is that while the idea of form is completely intuitive, it is not truly self-evident in the sense that it can be accepted without proof. In fact, one of the strange results of non-Euclidean geometry is that there are no similar triangles. The sum of the three angles of a large triangle will be less than 180 degrees, and they will also be less than a smaller triangle.

It would fall to another, later generation of mathematicians who were much more imaginative than Wallis to solve this mystery. Nobody had any idea how much geometry would change with the solution.

In the eighteenth century, interest in the fifth postulate really took off. Mathematicians all over Europe were hard at work on it. Georg Klügel in Germany, Johann Lambert in Switzerland, and Adrien-Marie Legendre in France all failed to prove it. Legendre tried and failed to solve it his whole life. Until the early nineteenth century, the best that anyone had ever done was to write a clever alternative way to express the fifth postulate. The most famous of these was Playfair's axiom, a restatement of the postulate, which John Playfair published in 1795. He wrote a definitive version of Euclid's *Elements* that was also a complete revision. His axiom states simply, "Through any point in space, there is at most one straight line parallel to a given straight line." Pretty, but no solution.

The interest in the fifth postulate paralleled a general growing interest in geometry as a whole. There were 84 new editions of Euclid's book in the sixteenth century, 92 in the seventeenth century, 118 in the eighteenth century, and more than 150 in the nineteenth. Dozens of German editions would come out during Gauss's lifetime, and from 1840 to 1879 there were nearly one hundred new editions of the *Elements* issued in England alone.

In the eighteenth century, it had also become trendy to rewrite geometry. Attempts to reform the teaching of the subject were not a reaction against Euclid but a reassessment of the book in light of the ability of European students of those days to learn. By the time Gauss was born, Euclid had ceased being taught as an elementary textbook. More common were other books, liberally sprinkled with examples from surveying, billiards, warfare, and other entirely practical applications. The mark of Euclid was everywhere present in these books. Euclidean geometry was the basis for the formulations, proofs, and fundamental concepts, but it was not the basis of the presentation. A few years before Gauss was born, one commentator even asked why anyone would bother with Euclid at all.

Some even speculated that teaching Euclid did as much harm as good. By the time Gauss was born, England was the only country in Europe where Euclid was still taught. The problem was that much of the new work did little to improve understanding of geometry. In 1784, the mathematician d'Alembert even complained that most elementary geometry texts were written by men of little ability. This soon changed the year Gauss started college, when Legendre published what would be perhaps the greatest book on geometry since Euclid. It was hugely successful, and various editions remained in popular use in European schools until the twentieth century. Legendre's 1794 book also attempted to prove the fifth postulate.

Göttingen was one of the centers of failed proofs, and almost as soon as Gauss arrived, he was thrust into an arena where interest in proving the fifth postulate was very high. The university took the unusual approach of making an official decree that the postulate must be true.

When Gauss arrived at the university, a professor had just published three proofs of the fifth postulate—all incorrect. The following year, another professor named Seyffer would publish two separate reviews of what other mathematicians had done to try to prove the fifth postulate. His conclusion was that it may be impossible to prove. Seyffer's opinion must have resonated loudly with the impressionable young Gauss, who was close to the professor.

As it turned out, Seyffer's idea that the fifth postulate could not be proven was the gateway to the discovery of non-Euclidean geometry. Gauss would soon cross that threshold, and this is where he and his college companion Farkas would part.

Like all gifted young mathematicians of their day, both Farkas and Gauss developed a common interest in proving the fifth postulate, but they followed different trajectories. By the end of the eighteenth century Gauss had been thinking about it for a few years and seems to have recognized the effort as futile, perhaps because it could not be proven. A lot of what we understand about what he knew then depends on what he said to his colleagues in letters years later and notes that he made in his notebooks that were not discovered until after he died. But one thing is clear: he abandoned trying to prove the fifth postulate.

Farkas, on the other hand, made it a lifelong ambition to find the proof. He never let it go. He again resorted to the tried-and-untrue proof that relied upon looking at straight lines that were everywhere equidistant. Eventually, he showed that the fifth postulate can be deduced from the postulate that three points that do not lie on the same line will always lie on a circle. It was an outstanding achievement—perhaps as far or further than anyone had ever gone before. But it didn't really matter. He still had not proven it, and he knew it.

"Forsoothe I wish the youth by my example warned, lest having attacked the labor of six thousand years, alone, they wear away life in seeking," Farkas wrote later in life. In the end, he and Gauss went their separate ways, Gauss back to Brunswick, and Farkas back to Hungary. In 1807, Gauss became a professor at the University of Göttingen and the director of the observatory. Farkas became a professor at the Reformed College of Maros-Vásárhely, where he remained for the next forty-seven years.

In the summer of 1799, the two men met for the last time in Clausthal, a village conveniently located halfway between Göttingen and Brunswick. "That feeling of seeing each other for the last time is indescribable. Even a word about tears is ineffective. The book of the future is closed," Farkas wrote of their meeting.

It was not their last communication, however. Farkas continued to work on proving the fifth postulate. He shared some of this work with Gauss in 1799, and Gauss wrote to him that same year, telling him to publish his work right away but warning him that he would receive little by way of recognition because there would be few who would be able to understand it.

Gauss understood the problem quite well. He even had thought of his own way to solve it, but by the end of 1799, he still had not done so. "Indeed I have come upon much, which with most no doubt would pass for a proof, but which in my eyes proves as good as *nothing*," he wrote to Farkas on December 16 of that year.

He concluded the letter by saying that he still was not satisfied. "As for me, I have already made some progress in my work," he wrote. "However the path I have chosen does not lead at all to the goal which we seek, and which you assure me you have reached. It seems rather to compel me to doubt the truth of geometry itself." Gauss was toying with non-Euclidean ideas, but he was not ready to accept them.

Farkas, on the other hand, seemed never to veer from the path of trying to solve the fifth postulate. After he became a professor of mathematics in Hungary in 1804, one of the first things he did was to write up a formal treatment of the work he did on the fifth postulate when he was in Göttingen. He must have felt like he almost had it, and in 1804 he sent Gauss his "Göttingen theory of parallels." In this, Farkas conceived of constructing something like an open box . . . or rather a rectangle with the top part missing. He attempted to show that the top part had to be parallel because if it weren't, it would lead to a contradiction. This was no proof, however, because it still did not start from an easily accepted premise that did not need to be proven.

Gauss was gracious to a degree rarely seen in any age. He once told his friend that if Farkas had solved the fifth postulate before him, he would do everything he could to advertise the accomplishment. But Gauss recognized right away that Farkas had not done so. He replied in a letter a few months later that he was delighted with Farkas's work and that much of it mirrored his own thinking. Recognizing that it was still not solved, he promised to revisit the problem at a later date, though he apologized that he could not do so then.

Gauss wrote, "I have read through your treatise with great interest and attention, and am delighted at its really profound keenness." It was a lot like what Gauss had himself done, so he knew it to be fruitless. "You may wish only my candid, open judgment," he wrote. "And this is that your procedure does not give me satisfaction."

He said there was still a stumbling block in that the basis for Farkas's proof could not be proven without first assuming the fifth postulate to be true. He pointed out that what Farkas had done was not a proof but yet another unproven postulate—no closer to the truth of geometry than two thousand years before.

Gauss told Farkas that he hoped that someday he or someone else would be able to solve the riddle. "Your method does not yet satisfy me," he wrote to his friend. "I still hope that these cliffs will be navigated eventually, and this, before I die." Then he apologized that he had a lot of pressing work and could not get to the solution himself. Probably Gauss was the more clever of the two and sensed that the proof would not be forthcoming. Farkas was less insightful and would spend the next two decades continuing to toil away at this frustrating problem too simple to forget but impossible to get over.

Gauss didn't correspond seriously with Farkas again until 1832. While interest in the fifth postulate was something they had shared for a few years, Gauss also grew to be much more as a mathematician. He was deeply interested in several other areas of the discipline, and for the next few years he allowed himself to be distracted by them. When he left Göttingen to return to Brunswick, he was satisfied to keep his radical ideas about non-Euclidean geometry completely to himself.

The end of the eighteenth century was tough on Gauss. When he left Göttingen in 1798 and returned to Brunswick, he had almost nothing to show the duke to justify his investment in his education. His sole publication was the one paper on the seventeen-sided figure, a significant but obscure and esoteric study. Gauss did not graduate from Göttingen but earned his Ph.D. from the University of Helmstadt in 1799 for his four proofs of the fundamental theory of algebra—a complicated mathematical theorem that describes algebraic equations and how they can be factored in terms of their complex roots.

Mathematicians had been trying to prove this theorem for centuries, and several had offered what they believed to be proofs. One of Gauss's four proofs alone might have been enough to earn his doctorate, but he didn't like to do anything that was not complete. In his dissertation, he not only offered his own complete multiple proofs, but he demonstrated that the earlier proofs were all incorrect. "Doctoral dissertations, even of the greatest scholars, rarely exhibit anything more than passing value," one of Gauss's biographers wrote more than a century later. "Not so with Gauss."

Gauss had been working on the fundamental theory of algebra since starting college, and he had devoted himself completely to the task. He had completed a book on the subject by 1796, his second year at Göttingen, and he was essentially finished with it two years later when he returned home. The duke was still supporting him, so he moved out of his parents' home and set about trying to get an expanded version of this work published. Gauss later claimed that this was the most exciting and productive time of his life.

The duke supplied the necessary funds to ensure the book's publication, but the printing was slow and protracted. It began in 1798 and wasn't finished until three years later. This didn't necessarily bother Gauss. He was devoted to absolute perfection in his work, and he was happy to see the book come out perfectly, if not speedily. His mathematical idols were Newton and Archimedes. Gauss once called Newton the master of all masters, and Newton was similar to Gauss in the sense that he had published very few works in his early years before his masterpiece, the *Principia.*

Newton's book was the greatest scientific work of the seventeenth century, and when it was published Newton was a middle-aged man. Now Gauss, barely twenty years old, was about to publish one of the greatest mathematical works since the *Principia* came out more than a century before. His book, called the *Disquisitiones Arithmeticae,* became a standard text that mathematicians studied throughout the nineteenth century. The work followed from that of some of Gauss's contemporaries, like Lagrange. Parts of it, though, are completely original, and in some places he points out the mistakes of some of the much older and wiser mathematicians who had been working on these same problems for decades.

The *Disquisitiones* contained a number of significant developments. It introduced the concept of determinants, a fundamental identity in matrix algebra. It also contained Gauss's proofs of the fundamental theorem of algebra. The book originally contained eight chapters, but Gauss published only seven because it was too expensive to include the last. This was a minor hitch, though, because the work was remarkably complete.

The *Disquisitiones* was the mathematical equivalent of great art. One scholar says that it "has never really been surpassed or supplanted." No mistakes have ever been found in the book, and Gauss once boasted that if another edition were to be printed, he would not change anything. That he was in his teens and early twenties when he wrote it is incredible. It could be likened to a college student shooting a movie as good as *Citizen Kane* using his cell phone.

The book sold better than expected—which is to say not well enough to mean any real cash for Gauss. He made almost no money from it. But the *Disquisitiones Arithmeticae* introduced him to all the great and famous mathematicians of his day, not as a young up-and-coming scholar—not even their respected peer—but their master. "Your *Disquisitiones* have with one stroke elevated you to the rank of the foremost mathematicians," Lagrange gushed in a letter to Gauss. He also referred to one of the discoveries in the book as "one of the most beautiful which has been made in a long time." Laplace, another great French mathematician of the day, piled public praise upon Gauss. "The Duke of Brunswick has discovered more in his country than a planet," he said once. "[But] a supernatural spirit in a human body."

Gauss was exceedingly grateful to his patron, and in the dedication of his greatest work he thanked the Duke of Brunswick for his wise and liberal support. For his part, the duke gave Gauss a nice stipend and a free apartment. Not everyone immediately gave Gauss his due for this book. The book was introduced to the French Academy of Sciences by Legendre, and in its January 1802 minutes, the Academy notes the "geometric discovery of Charles Frédéric Bruce"—misspelling his name for the ages.

For the rest of his life, Gauss considered number theory the queen of mathematics. He continued to think about it even when his career took him in different directions, toward astronomy and physics. When the *Disquisitiones* was finally finished in 1801, its appearance would coincide with Gauss's meteoric rise to fame. The book was not the main reason for his rapid ascent, however.

The real reason for Gauss's fame was that many astronomers in Europe were desperately searching the skies for a lost planet. Gauss would find it, taking the opportunity to show the astronomers of Europe that he was someone whose genius would bear fruit in more than just number theory—a field that was likely as uninteresting to most of them as it was inaccessible. This would change Gauss's life forever.

6

Searching for Ceres

Ceres Ferdinandea,
Spherical king of the asteroid belt,
Long ago the missing planet,
A dwarf, discovered by a giant.

—Jason Socrates Bardi

One of the greatest discoveries of the nineteenth century was also the earliest. It was just after midnight on New Year's Day 1801, and an Italian monk and astronomer named Giuseppe Piazzi was up late searching the skies from his rooftop observatory in Palermo, Sicily. He saw something that night that would change his life forever, and Carl Gauss's life, too. Piazzi, peering through his telescope, spotted an object. He watched it for several nights, carefully recording its location as it traversed the sky.

Piazzi wasn't sure what he was looking at. Was it a planet? Was it a comet? His letters indicated that he thought it could be either—perhaps a comet without a tail. In one of these letters, Piazzi remarked that it might be "something better than a comet." What exactly was it? Perhaps it was what Piazzi was hoping to find: a new planet.

Finding the "missing" planet was, in fact, one of the great scientific quests of the day. It was a completely new and exciting endeavor that had started a generation before by the discovery of the planet Uranus. The ancients knew of only five planets other than Earth—Mercury, Venus, Mars, Jupiter, and Saturn. But the scientific world had been turned on its head when in 1781 William Herschel at his private observatory in Bath, England, spotted something he thought could be a star or a comet. It turned out to be the new planet Uranus.

This raised the possibility that there might be even more un-discovered planets in the solar system. J. D. Titus and J. E. Bode gave this possibility a firm theoretical basis by forwarding an idea that there logically must be a planet between Jupiter and Mars. The basis of their claim was their observation that the six known planets followed an orderly, even spacing, thus suggesting that all the planets should be so spaced. But the gap between Mars and Jupiter was about twice what would be predicted by their theory. The emptiness was deafening.

In a way, the concept went all the way back to Pythagoras. He held that different strings gave different notes based on their length while at the same time different planets find different notes based on their distance from the sun. The missing planet was like a missing string on an instrument.

By the turn of the eighteenth century, astronomers had been looking for a planet between Mars and Jupiter for several years, and the search was heating up. Baron Franz Xaver von Zach, the court astronomer at Gotha, began a formal, systematic search involving a team of astronomers. Piazzi, in his observatory in Sicily, was one of a group of twenty-four astronomers in Europe who made up this team. With such a large force of astronomers gearing up to look for the missing planet, Piazzi must have been surprised when he hit pay dirt almost instantly that New Year's Day, spotting an object that appeared to be in orbit around the sun between Mars and Jupiter, exactly where the missing planet should have been.

Piazzi continued to track the object as long as he could, but he became ill on February 11, presumably from too many sleepless January nights exposed to the elements. In any case, the object was about to pass behind the sun, and he soon lost sight of it. By then, Piazzi had decided to presumptively name his object. He called it Cerere Ferdinandea, combining the names of a Roman goddess of the harvest with his own Sicilian king. We know this object by its more common name: Ceres.

Piazzi had already sent letters that January to the directors of the observatories in Paris, Berlin, and Milan, telling them that he had discovered a small comet with no tail—or something—and giving them the coordinates. In one of the letters, he was bold enough to suggest that Ceres might be a planet.

The letters, couriered via the unreliable and slow communications of the day, took a long time to reach the observatories. The letter to Berlin, for instance, took nearly three months. But by that summer, word had gotten around. The discovery was hailed in a long article in a German magazine called *Monatliche Correspondenz*, edited by Baron von Zach. The magazine echoed the sentiments of a generation of astronomers who could not have been more excited by the letter from Piazzi. The description of Ceres presupposed that it was a new planet.

This was "a long supposed, now probably discovered, new major planet of our solar system between Mars and Jupiter," wrote Baron von Zach, who was one of the most famous astronomers in Europe at the time. Von Zach was the center of German astronomy. His observatory was the most modern in Germany, and the *Monatliche Correspondenz* was the nation's most important astronomy journal.

As confidently as von Zach wrote in his journal, he still had very little to go on. For the discovery to be meaningful, this mysterious new planet had to be located again. The only problem was, nobody knew exactly where to find Ceres. It had disappeared. This started one of the great hunts in the history of astronomy. Astronomers, professional and amateur alike, began scanning the skies with telescopes to locate the mysterious new rock, without success. Ceres was still there, of course, rounding its way around the sun far beyond the orbit of Earth. And all the astronomers tipped off to its existence knew it must be out there—but where?

This may have been a great challenge in astronomy, but it was also a wonderful problem in applied mathematics. Mathematics alone could (and eventually would) locate the planet without the aid of a telescope if it could be used to compute Ceres's orbit. If the astronomers of the day were given the orbit of Ceres, then it would be a simple matter for them to locate the missing planet by searching along the path where they already knew it should be. But how could mathematicians determine the orbit?

Perhaps the easiest way, in theory, would be to chart it—making careful observations of its exact location in the sky night after night as it made its journey around the sun. If one were able to chart all the locations on a solar system map accurately, it would be a simple matter to connect the dots.

For an object in Ceres's orbit, this would take about five years, assuming that you could observe it continuously. Of course, things are not always that easy in reality. At certain times of the year, the object would be behind the sun or in the wrong position relative to Earth and not observable at night. Add to this the difficulty that planetary motion in the heavens is not due to the movement of the planet alone. We do not enjoy a stationary vantage point but rather observe the motion of celestial objects from our platform of Earth, which is moving relative to the other objects in the heavens.

Nevertheless, if you were able to observe enough positions of an object like Ceres on enough nights, and if you were able to record the observations, you would have enough of a plot of its actual orbit to be able to fill in the gaps.

What if all you had was enough data to fill one tiny gap? How could you determine the orbit of an object like Ceres if there was only a limited amount of available orbital data? This was exactly the situation in which mathematicians and astronomers found themselves in 1801. Piazzi had carefully recorded the position of Ceres on several nights over a period of forty-one days from his observatory in Sicily. These observations could be plotted as points on Ceres's orbit, but the data was very limited.

Many astronomers of the day reasoned that it would still be possible to calculate the entire orbit from these observations. Suddenly the problem was clear. With enough mathematical analysis, and given the several known data points on the route of the rock around the sun, they could estimate its orbit, and from that they could determine where Ceres should be by calculating how far it would have traveled since Piazzi last saw it. One would then have to simply point a telescope to the location farther along this path where it should be. What a delightful solution in applied mathematics!

With that in mind, an astronomer named Johann Karl Burckhardt published a preliminary orbit of Ceres in July, and in September von Zach published the complete set of observations made by Piazzi. They were both hopeful that this would be enough. It was not.

Burckhardt's tools were relatively new, historically speaking. The mathematics that allowed him to attempt to determine orbits would have been completely unavailable to astronomers of ancient times. But in the previous two centuries, mathematics had brought new powers to astronomy, and a succession of astronomers would forever change the way that humans see their place in the universe—something that inspired the mathematician Laplace to call astronomy "the most beautiful monument of the human mind, the noblest record of its intelligence."

This development was hard fought every step of the way. Sometimes there were religious and intellectual authorities in the way—people who found mathematical discoveries hard to swallow. Galileo endured criticism for his discovery of the moons of Jupiter, for finding sunspots, and for his support of Copernicus, because the commonly held belief was that the sun was a perfect orb and that it, like all the heavenly bodies, revolved around Earth. Jupiter's moons drew criticism because they increased the number of bodies in the solar system from seven, which was thought to be the logical number.

In the end, Galileo had to renounce his discoveries. "I abjure, curse, and detest the said errors and heresies and generally every error and sect contrary to the said Holy Church; and I swear that I will nevermore in future say or assert anything verbally or in writing which may give rise to a similar suspicion in me," he attested. Pity poor Galileo for having to retract what he knew to be right. The intricate and irregular motion of the planets was something that thinkers had struggled to explain satisfactorily, and the heliocentric solar system was a way of doing just that.

Sometimes the difficulty in astronomy arose due to the complicated nature of the problem. Johannes Kepler derived the basic laws that described orbital motion, and Newton worked out the mathematical framework, explaining planetary motion as a consequence of gravitational attraction. Mathematicians of the eighteenth century fully developed Newton's foundation and worked out the tools necessary to solve the problem of calculating orbits. Euler, Clairaut, Lagrange, and Laplace adapted powerful tools for planetary dynamics, and by the time Gauss was in university, these tools—called celestial mechanics—were determined, published, and understood.

By the end of the eighteenth century, the proper physical theory of gravity and planetary motions was in place, and a powerful calculus-based analysis was developed enough to deal with it. The theories had also already been successfully applied. Even so, the problem of determining an orbit from a few observations was an exceedingly complicated one, as Newton himself had admitted, and things were no better a century later when astronomers were trying to deal with Ceres.

The difficulty was that prior to the discovery of Ceres celestial mechanics had only been successfully applied to predicting the orbits of comets. Comets are much brighter than planetary bodies and may be visible for a longer period of time on their approach to the sun. They are easier to handle, in other words.

When Uranus had been discovered, it was near the farthest point of its orbit. This was a lucky break because it simplified the calculation. But what could be said of Ceres when it was observed? Was it near its closest point, its perihelion, or near its farthest point, the apihelion? Knowing these details could make a profound difference in the shape of the calculated orbit and thus the accuracy of the prediction.

Ceres is not bright enough to be visible to the naked eye, so it had to be spotted with a telescope. This greatly limited the area of the sky that could be systematically searched. And if astronomers did not know where to look, they were not going to spot it. So they needed the orbit to be predicted if they were ever going to find Ceres.

Computing the orbit of Ceres was exceedingly difficult because the observations Piazza made were but a tiny section of Ceres's arc. Recording observations taken over just a few weeks only gives you a few closely spaced points on the overall path of the object around the sun. Piazzi could observe only nine degrees of arc of the asteroid's orbit. Such a small piece of the orbit confounded any attempt to extrapolate the entire orbit. Seven different elements of the object's motion had to be computed to determine the orbit. With only a few observations, small errors in the observations could have massive effects on the overall calculation.

This is exactly what happened to Burckhardt's calculated orbit. A month after Baron von Zach published the full set of observations from the astronomer Piazzi, the baron also published the sad news that several astronomers had been looking for Ceres in the previous two months without success. Burckhardt's efforts to plot the orbit had failed.

This was the desperate state of affairs Piazzi, von Zach, and the others were in when the data in the *Monatliche Correspondenz* reached Gauss, sometime in the fall of 1801. Gauss was not an astronomer in the regular early-nineteenth-century sense. He had neither telescope nor observatory to follow planets around the corners of the solar system. But, being adept at pushing pencils around on a desk, he was perfectly suited to tackle this problem. He had already studied the moon's orbit on paper, and he had been examining the mathematical theory behind orbital calculations.

In 1801, Gauss was already studying the work of the great French mathematicians Lagrange and Laplace, which he had just obtained. These works were the culmination of a century of mathematical physics that married calculus—the great mathematical invention of Newton and Gottfried Wilhelm Leibniz—with the physical foundation that Newton had laid out in the *Principia*. When Ceres swept across Gauss's radar, he pushed everything else aside, picked up his pencil, and went to work. He then did what no other astronomer who tried to tackle the problem could.

"Could I ever have found a more seasonable opportunity to test the practical value of my conceptions, that now employing them for the determination of the orbit of the planet Ceres, which after the lapse of a year must be looked for in a region off the heavens very remote from that in which it was last seen?" Gauss wrote. Within weeks, he had found the answer, inventing along the way the mathematics necessary to do so. By November 1801, he began making notes on the problem, and by the end of the month he had solved it, working out a solution to a system of seventeen equations that predicted where Ceres would be. He communicated his calculated orbit to von Zach, and in December von Zach published the predicted orbit, writing, "Great hope for help and facilitation is accorded to us by the recently shared investigation and calculation of Dr. Gauss in Brunswick."

By then, astronomers were already searching the skies for Ceres, looking where Gauss predicted it would be. Winter weather being what it was, they were not immediately successful. On December 7, however, von Zach located the planet almost exactly where Gauss predicted it would be, as Gauss himself wrote, on "the first clear night, when the planet was sought for as directed by the numbers deduced from it, restored the fugitive to observation." A little later, a physician and astronomer named Heinrich Olbers also spotted the planet—also exactly where it was supposed to be according to Gauss.

Olbers was deeply impressed with Gauss, which says a lot for someone as accomplished as he was. Olbers was almost twenty years older than Gauss and had also studied at the University of Göttingen, but in medicine. He was a practicing physician, but his lifelong love was for astronomy, and the upper floor of his house was outfitted as an observatory. He spent a considerable amount of his free time peering at the heavens. He is said to have worked all day in his medical practice and then spent most of the night looking through his telescopes, sparing himself but a few hours for sleep.

The year Gauss was born, Olbers was a nineteen-year-old budding intellectual, and he made careful observations of a solar eclipse. Two years after that, he calculated the orbit of a comet that had appeared, a feat he would try to repeat with Ceres but failed. To find himself surpassed by a kid who had attended the same university—needless to say, he was deeply impressed with the young man. The two became lifelong friends and correspondents after that.

Olbers was not the only admirer of Gauss. From New Year's Eve 1801, exactly one year from the time Ceres was first spotted by Piazzi, astronomers all over Europe began to find Ceres and follow it. Many of them rang in the new year peering through a scoped lens, and probably many of them raised a glass to Gauss in celebration of his miraculous achievement.

The secret to Gauss's success was a technique he invented called the method of least squares, which is a way of minimizing the error of observations. The problem, as Gauss saw it, was to determine the orbit by finding a curve that corresponded with the observed data and then correct the curve to find the best fit. The method of least squares helps by approximating an orbit based on the few observations and then improving, or "fitting," the orbit to the data. The fitting works by minimizing the differences between the observed and computed points in the orbit multiplied by themselves (thus "least squares"). It was a simple but remarkable discovery and perhaps the single most valuable tool ever devised for analyzing data of any sort, even to this day.

The foundations of Gauss's methods had been known for a long time by the time he arrived on the scene, but nobody had ever put the technique together the way he had. There were no general approaches such as his. Several prior mathematicians had addressed the issue of correcting observations by finding a best fit. Laplace, in particular, had worked on similar applications. He had come close, but Gauss was the first to work out the method to its fullest development.

Gauss discovered the method of least squares in 1794, when he was still in university—a remarkable achievement for someone who was then just a teenager. "Few branches of science owe so large a proportion of subject matter to one man," one of his biographers wrote. The truly remarkable thing about Gauss, though, is that when he invented the method of least squares, whatever he felt about the utility of the method, he didn't think much of it. He didn't even bother to publish it because it was so obvious to him that he was convinced some other mathematician had surely thought of it already! In fact, other mathematicians did develop the method of least squares independent of Gauss, including the French mathematician Adrien-Marie Legendre, who was also the first to publish the method and was subsequently recognized as its inventor.

This was one of the peculiarities about Gauss. He didn't always publish his work, and much of the same ground was later covered by other mathematicians independently. In his early days of working on theoretical astronomy, he studied complicated functions known as elliptical integrals, which he never published. When the mathematician Abel later published a paper independently duplicating Gauss's work, Gauss was relieved because it meant that he didn't have to do so himself.

Gauss was overjoyed about the rediscovery of Ceres, and his enthusiasm was echoed by dozens of his contemporaries, many of whom had been trying to do just what he had done. In his overly humble way, he was careful not to promote himself too aggressively, and he credited much older mathematicians like Isaac Newton for working out the theoretical foundations upon which he laid his mathematical prediction.

Nevertheless, the calculation launched Gauss to fame and elevated him to a stature on a par with the top astronomers in Europe. After locating Ceres, he was considered one of the most respected astronomers in Europe. When Alexander von Humboldt returned to Europe from the United States in 1804, he went immediately to Paris, cosmopolitan capital that it was. There he was impressed to find the name of a young German mathematician mentioned over and over as one of the great geniuses of the day. When the king of

Prussia invited Humboldt into the Berlin Academy of Sciences, he responded that the only man who could elevate the academy's ranks was Gauss.

Gauss's rediscovery of Ceres had a profound effect on his personal life. He began to spend more and more time on astronomical observations, carefully following planets and comets and watching eclipses. On many nights during the next half century, he could be found late at night observing the stars through his telescope and taking measurements with his sextant and recording all. He wrote to his friend Farkas Bolyai on June 20, 1803, that astronomy and mathematics were the two "magnetic poles towards which the compass of my mind ever turns." He also named his firstborn son Joseph, after Piazzi.

The success launched Gauss's career as well. He soon gained membership in numerous scientific societies, and by the time he died, he would be a member of all the major societies in Europe. The Royal Society of London elected him a foreign member in 1804. The Academy of Sciences at St. Petersburg elected him a corresponding member. A very high minister in the Russian government named Nikolaus von Fuss wrote to Gauss personally to offer the good news. In 1802, the czar of Russia himself attempted to lure Gauss to serve as the astronomical director at the observatory at the St. Petersburg Academy. This was followed in 1807 with an offer to come and lead the observatory at St. Petersburg. Gauss was also given several offers to take up posts elsewhere, including at the University of Göttingen, where he was offered the job as the director of the astronomical observatory.

Several months after Gauss's great discovery, however, the Duke of Brunswick, his longtime patron, awarded him a nice annual salary. When he was told about the award, Gauss is said to have exclaimed, "But I certainly haven't earned it. I haven't done anything for the country yet." Almost two years to the day that Ceres was rediscovered exactly where Gauss had predicted it would be, the duke generously increased his salary again and added a housing allowance and an annual allotment of firewood. All of this was given to him without any responsibilities in return.

Gauss subsequently set about writing down the methods that had led him to the discovery, in the book *Theoria Motus Corporum*

(Theory of Motion of the Heavenly Bodies). This was the perfect marriage of theory and practice, and it became the most influential astronomical text for several decades. The Napoleonic Wars disturbed things so much in Germany that Gauss was almost unable to find a publisher for his book when he finished it in 1809, but he was so sure of his talent that he begged the publisher not to cut corners on its production, with the promise that it was such an important book it would "still be studied after centuries."

The book was not without controversy. Because it arrived a few years after Legendre had published a method of least squares, some (like Legendre) assumed that Gauss borrowed the technique from him. Nowhere in the book did Gauss credit Legendre, however. This wasn't much of an issue for Gauss. He didn't regard it as his greatest discovery, but he definitely thought of it as his own work. Because he viewed it as his own discovery, he saw no need to credit Legendre.

Legendre was not satisfied with such explanations. "How can Mr. Gauss have dared to tell you that the greater part of your theorems were known to him and that he discovered then as early as 1808?" the Frenchman raged in a letter to a mutual friend. "This extreme impertinence is incredible on the part of a man who has sufficient personal merit to have no need of appropriating the discoveries of others."

For his part, Gauss was not impressed. "I had no idea that Mr. Legendre would have been capable of attaching so much value to an idea so simple that, rather than being astonished that it had not been thought of a hundred years ago, he should feel annoyed at my saying that I had used it before he did," he wrote.

Controversy aside, the book brought Gauss added fame. In 1810, the Institute of France awarded him a prestigious prize based on it, but Gauss refused the money. He did, however, accept a nice pendulum clock, which he kept for the rest of his life. He had come a long way from the boy who would study by the light of a turnip filled with animal fat.

Upon his newfound reputation shortly after his discovery of Ceres, Gauss could have seized the moment to publish something—anything—about non-Euclidean geometry. There may have been little that he could have tackled or published that would not at least have been considered by his contemporaries. But his reputation was certainly not unassailable. Science in the beginning of the nineteenth century did not hold the lofty position in society that it does today. Even so, respect for him was solid enough that he might have questioned the fifth postulate without facing the terrible criticisms he may have feared.

Gauss was well aware, however, of the pettiness to which some stooped in scrutinizing scientific discovery. He once wrote to von Zach, upon hearing of criticisms that the discovery and orbital calculation of Ceres was an utter waste of time, "It is scarcely comprehensible how men of honor, priests of science, can reveal themselves in such a light."

7

The Dim Light
of Exhaustion

We cannot confuse what to us appears unnatural with the absolutely impossible.

—Carl Friedrich Gauss

Carl Gauss saw himself as a historical figure—even at an early age. By the time he was in his twenties, and especially after his discovery of Ceres, he knew that his place in the history books was assured. He was certain his work would be highly regarded by future ages of mathematicians.

None of this seems to have ever gone to his head. Perhaps it was his humble upbringing, maybe some innate strength of character, or perhaps even it was the dangerous Napoleonic times in which he lived (he never ventured to stick his neck out very far). Whatever the origins of his personality, Gauss was a thoroughly staid soul— not the sort of man who would invite controversy of any sort. The first words his greatest biographer used to describe him in the very first paragraph on the first page of his biography were "not controversial."

Gauss was always cautious in publishing anything. He could have described some of his work on imaginary numbers in his doctoral thesis in 1799, but he was much too wary to do so, and he waited more than three decades before publishing the work. This is not to say that he was a scholarly wallflower. He had stiff competition and he knew it. He wrote to his friend Heinrich Olbers in 1806 that he seemed to be competing with Adrien-Marie Legendre on everything he worked on. This was not a competition that Gauss shied away

from. He was very critical of Legendre, in fact, and this was probably
a large part of the reason Legendre had no love for him.

But it was not concern over brilliant mathematicians like
Legendre that explains why at the turn of the century Gauss didn't
publish anything about non-Euclidean geometry. He was more concerned about the outcry of the lesser mathematicians in response
to anything he might publish. Gauss saw most mathematicians—
indeed, most people—as lacking the rigor needed to come to understand and appreciate generally established truths in ordinary mathematics. This placed new mathematical discoveries far outside of
their reach. Gauss had written to his friend and student Christian
Ludwig Gerling in 1815, "Even our greatest mathematicians have
mostly somewhat dull feelers in this respect."

Non-Euclidean geometry would raise the bar even higher because it would require people to accept strange concepts like the
idea that straight lines could be curved. Perhaps Gauss chose not
to publish anything because he was aware of the complicated nature
of the subject.

Besides that, Gauss also had the sort of constitution that forbade
him from making his inventions public before they were completely
ready to be communicated. This tendency was not out of any desire
to hide his results, but rather the need to first thoroughly complete
his works. He did not like to even discuss his views in public unless
he deemed them ready for publication. Sometimes even then he did
not publish them. When he was an old man, for instance, he once
corresponded with a much younger Carl Gustav Jacob Jacobi, who
had done some work on elliptic functions. Gauss had become aware
of them and told Jacobi that he had developed the same results
years before.

Only those ideas that were the most developed were *ever* that
ready. Gauss's motto in publishing was "few but ripe," or, in Latin,
pauca sed matura. "I have many things in my papers, which I may
perhaps lose the chance of being first to publish; but so be it, I pre-
fer to let things ripen," he wrote.

Gauss was also enjoying a relatively stable existence—despite the
fact that the Napoleonic campaigns were about to bring war to his
front door and throw his future into uncertainty. He was already
famous and recognized far and wide for his genius. He enjoyed
a slightly improved social position, and his few needs were met
through the beneficence of the duke. All this greatly enhanced his

personal intellectual freedom. Mathematics now reigned supreme in his mind. Later, when he was an old man, enjoying the unabashed venerable fame that comes only with age and genius, he would often reflect fondly on these years in the early nineteenth century.

These were the happiest times of Gauss's life.

Gauss's greatest joy came on October 5, 1805, when he married his sweetheart, a plain, loving girl named Johanna Elisabeth Rosina Osthoff. She was the daughter of a leather tanner, and it is fitting that Gauss wound up marrying a tanner's daughter. When he was a boy, he had an intimate connection to the tanning trade through his mother. Before Gauss's mother married, she was employed by a tanner named Mr. Ritter, who may have been the same person as Friedrich Behrend Ritter or George Karl Ritter, who were both god-father to Gauss. Coincidentally, Johanna's father was also named Ritter. To Gauss, Johanna was an angel. He called her Madonna and gladly devoted his life to her.

The two met in 1803, when Gauss was basking in his newfound fame, and he fell in love with her from the moment they met. The way he talks about her in letters is so romantic and true to his feelings as to be inspiring—even when translated from his eloquent German, which has been said to be masterful. Gauss described her in a letter to Farkas Bolyai as "exactly such a girl as I have always desired for a life companion."

Johanna was very interested in Gauss, too, but she had heard that he was romantically involved with a wealthy young local lady, so their courtship was delayed by several months. Eventually, though, Gauss convinced Johanna of his true feelings, which he spilled out in a letter he wrote to her in the summer of 1804: "My heart knows your worth—O! more than it can bear with ease. For a long time it has belonged to you. . . . Can you give me yours?" He promised her a life and "a true heart full of the warmest love for you," and con-cluded the letter of proposal with yet more romantic overtures. "Dearest, I have exposed to you my inner heart. Passionately and in suspense am I waiting for your decision. With all my heart, Yours, C. F. Gauss."

They were engaged in November 1804. Gauss wrote to Farkas a few days later, "Life stands like an everlasting spring with glit-tering new colors before me. . . . For three days that angel, almost too celestial for this Earth, has been my fiancée. I am superbly happy.''

Farkas warned Gauss not to trust women at all—even those who appeared pure and crystalline in their beauty. Recalled Gauss, Farkas's exact words were, ''The white snow passes away and leaves after it a black muddy filth.''

This did little to dampen Gauss's spirit. The following October he and Johanna were married, and their first son arrived on August 21, 1806. At the time, Gauss was beginning an effort to build an astronomical observatory at Brunswick, and personally and professionally he could not have been happier.

All that lasted just a few months. From 1805 to 1815 France would wage almost continuous warfare. Tranquillity in those years was fleeting. In 1802, France received the Louisiana Territory from Spain as a spoil of war and turned right around and sold it to the Americans. Napoleon had an almost unquenchable thirst for power. It was largely his ambition that would drive the Napoleonic Wars in the next decade.

Gauss's benefactor, the duke of Brunswick, was mired in the politics of Europe, and he traveled to St. Petersburg in 1805 to seek an alliance with the Russians against the French. Russia refused. While the duke was in Russia, he was pressured to release Gauss from his bond so that he could leave Germany and come to set up shop in St. Petersburg. The trip was wasted from both sides. Gauss didn't go to Russia, and as soon as the duke returned to Germany, he raised Gauss's salary.

Gauss did not get to enjoy this raise for very long. In 1806, the duke led a disaster of an effort to confront Napoleon. Duke Ferdinand was an accomplished soldier with a career stretching back to his days as a young officer in service of Frederick the Great. In recognition of his distinguished career and great ability, he was selected to lead the forces against Napoleon, and he was made generalissimo of the Prussian army. Prussia, renowned for its mighty army, seemed a formidable force to lead a coalition against France, and Duke Ferdinand was a force in his own right. Nobody could have expected that things would go as they did.

His army was poorly trained, ill-equipped, paralyzed by incompetence and infighting, and sorely outnumbered by the French.

Napoleon met the Prussian forces at Auerstedt and Jena on the morning of October 14, 1806, crushed them, and quickly occupied Berlin. The first battle of the war was a rout.

Gauss's patron suffered a grisly fate. The duke, overworked and exhausted, led his troops into battle on the foggy autumn morning and was shot in the eye and struck blind. The musket ball shattered his nose and gave him a horrifying appearance for his last days on earth.

About ten days after the injury, a mournful, slow wagon train exited the city gates of Brunswick at dawn. Gauss woke early that morning to the sound of a train of coaches leaving town. It was his patron the duke fleeing on his back. Gauss watched in sadness as his friend and benefactor was carried away in a last desperate attempt to escape. It was not to be. The duke's injuries overwhelmed the modest medical capacities of early-nineteenth-century physicians. A few weeks later, the duke died. He breathed his last breath of feudal Europe, and after he died, the old Europe Gauss knew in his youth passed away as well.

Napoleon organized the Confederation of the Rhine in 1806, which ended the Holy Roman Empire. Of all the profound political differences between that time and now, perhaps the greatest and strangest was the existence of the Holy Roman Empire. It had survived a thousand years, through countless and sometimes almost continuous warfare in the previous few centuries—not to mention the Protestant revolution, which divided the German kingdoms. Now it was dissolved, and the kingdom of Westphalia, incompetent and doomed, stood in its place over Brunswick and Gauss.

Though just shy of thirty years old when his patron died, Gauss had already achieved greater success and recognition than most scholars could hope for over their entire careers. The French occupation of Brunswick threw his career and his life into turmoil. Gauss was not sure of his next move, when an officer in the French army came knocking on his door. This officer had been dispatched from a French general in Germany who had come from Paris after having received personal entreaties to protect Gauss. The general had been told that Gauss was a good person and worthy of sparing, and he had been put up to this by a young woman in Paris named Sophie

Germain—someone Gauss had never met but with whom he had corresponded over his *Disquisitiones.*

The *Disquisitiones* was complicated enough to be almost incomprehensible, so Gauss had been excited when he started receiving letters from a young man in Paris who called himself Le Blanc. Gauss was impressed that this Le Blanc had not only read but understood the book. He wrote to Olbers, his friend the doctor and astronomer, that a mathematician had been communicating with him from Paris. The excitement that Gauss felt was surely that of a first-time author who had suddenly received praise and adulation from a complete stranger. The mysterious Le Blanc turned out to be Sophie Germain.

Sophie came from a good family. She was a teenager in Paris when the French Revolution broke out, and she found refuge in those dangerous and uncertain times in her father's library. There she acquainted herself with enough Latin to be able to read the works of Isaac Newton and other mathematicians. She had some of her mathematical papers smuggled to Legendre, who immediately recognized the brilliance of the author. Discovering the writer to be a young lady, Legendre became her private mathematics adviser. Sophie became somewhat celebrated for her genius—both because she was a woman and also because she was so young. Into adulthood, she continued to use the name Le Blanc, however, and this was how she wrote to Gauss.

A year older than Gauss, Sophie had spent much of her childhood sneaking time to study mathematics, much as Gauss had done. As a young girl, she was enamored of a story about the death of Archimedes. The old, overworked mathematician was devising machines and devices to defend his native Syracuse against the assaulting forces of Rome, and the city was breached by betrayal and fell to the Romans. Archimedes was so wrapped up in working out some mathematical formula that he barely even noticed the Roman soldier who burst in and killed him.

This story struck Sophie so soundly that she became resolved to understand mathematics. She would study day and night. Her parents, alarmed at her interest in what they considered an unladylike endeavor, started putting out her fire and taking away her clothes at night in the hope that this would force her to go to bed. Instead, she would wrap herself in blankets and study away. It would get so cold that the ink she used to write with would freeze on the nightstand. She succeeded wildly, teaching herself all of the mathematics of her day and becoming one of the greatest mathematicians of her era—not to mention a heroine of sorts for interceding on behalf of Gauss.

Gauss left Brunswick and moved to Göttingen in 1807—almost exactly a year to the day after the duke's sad demise. Göttingen was a city in transition in 1807. The old government was gone, and in its place was what would become the Westphalian government, an ill-fated and short-lived puppet regime set up by the French. When Gauss arrived, neither government was operating.

Gauss had people looking out for him there as well. French forces were sweeping through Germany, and Laplace asked Napoleon to exercise special precautions with Göttingen. "The foremost mathematician of his time lives there," Laplace told Napoleon. Had Gauss been more of a social climber, he might have used his credibility among the mathematicians in France to greatly improve his position. Napoleon appreciated mathematics and the mathematicians of his realm, and one of those whom he most admired was Laplace. Laplace in turn held an almost unique admiration for Gauss, and Gauss was certainly aware of this.

The taxing times must have weighed heavily on Gauss's mind, though. Napoleon was certainly one to enjoy the spoils of war. In the 1790s he invaded Italy and carried a fortune in money and artwork back to France. Now the rumor was that he was planning to sell Göttingen and the surrounding area to a nearby ruler.

Napoleon was not beyond such an act. He had already won and then turned around and sold a wide swath of Spanish holdings in the New World to Thomas Jefferson and the Americans—an act that U.S. schoolchildren would learn to call the Louisiana Purchase. The lesson of the exchange as it is often taught in the United States is one that unfairly favors the wisdom of the purchase. The United States got all those acres for pennies apiece. But for Napoleon, it was essential cash, money that he could use for his future war efforts.

Napoleon levied taxes on his subjects and raised money through other mechanisms as well, including compulsory loans by established institutions in Germany to the newly formed Westphalian government. One of these was the University of Göttingen, where

Gauss had just arrived. The burden of the massive loan demanded of the university was great, and Gauss was responsible for a large amount himself, even though he had yet to receive a single paycheck from his new employer.

Soon after the new government was put in place, it levied a war tax on its subjects. Gauss was liable for a considerable sum. Several people, including Lagrange in Paris, offered to pay it for him, but Gauss refused the charity. Still, he had to come up with a lot of money. Without hesitation, his friend Olbers sent him the cash, but Gauss refused to accept it. He sent the money back. In the end, Laplace paid the fine anonymously from Paris. When Gauss found this out, he eventually repaid the sum with interest.

French occupation brought a few good things. For some towns it was the first secular form of government since Roman rule. In other places, French administration was simply better. It introduced civic improvements that the population might not have even dreamed of—things like vaccinations and streetlamps. Most of all, the French occupation introduced the Napoleonic Code throughout occupied Europe. This set of laws helped to advance ordinary people throughout Europe by promoting civil equality and individualism.

Whether because Europe had a taste of the code's benefits or whether it was because of policies put in place following Napoleon's defeat, the continent enjoyed a period of relative tranquillity. There were no major confrontations for decades.

Prior to Waterloo, the 1815 battle that ended Napoleon's career, Gauss again began thinking about geometry, even toying with certain ideas related to non-Euclidean geometry. Somewhere in the midst of all the political chaos in the first decade of the nineteenth century, Gauss saw clearly the problem with the fifth postulate. Any proof was destined to fail because the fifth postulate could not be proven. It didn't need to be proven at all if it was not true. Gauss was well aware of an interesting approach that a few had tried in the eighteenth century. This was to deny the fifth postulate and examine what geometry would be like without it, showing that geometry without the fifth—non-Euclidean geometry—didn't make sense. One of the greatest conceptual breakthroughs of Gauss's life was realizing that none of that mattered. Geometry did not need to make sense to work.

It is hard to say when Gauss realized this exactly. His notes are rather short on specifics. Later in life he claimed to have had some sort of insight into non-Euclidean geometry as early as 1792. Evidence from his writing suggests that it could have been no later than midway through the 1810s.

In any case, perhaps prior to his realization that non-Euclidean geometry is perfectly valid, Gauss still attempted to prove the fifth postulate by proving that assuming it is not valid leads to a contradiction. He was developing this in October 1805, having worked on and off in the subject for a few years. How far he had come by 1808 is not clear. When given a chance to discuss the subject with an old friend, he declined.

Farkas Bolyai had a breakthrough—or so he thought—on Christmas night 1808. He sketched out his ideas in haste the next day, and the day after that he wrote a letter to Gauss telling him about his idea and asking for his opinion on it. "As soon as you can, write me your real judgment," he said.

Gauss demurred, apologizing that he was preoccupied with his own work, but he could barely contain expressing his domestic bliss to his friend. "The days go happily by in the uniform course of our domestic life," he wrote. "When the girl gets a new tooth or the boy learns some new words, then this is nearly as important as the discovery of a new star or of a new truth."

The solution to the fifth postulate never seemed so close to being within grasp. Now within reach after all those thousands of years, it would have to wait, however. Gauss was probably still not sure enough to accept non-Euclidean geometry. How could he? He saw geometry as an experimental science and not a purely theoretical one like number theory and the higher algebra that he truly loved. And Euclidean geometry—illogically footed as it may be—worked.

8

Gauss's Little Secret

Certainly the world did not hear of it from Gauss.
—George Bruce Halsted, an American mathematician

T he secret to solving the fifth postulate was not to prove it but rather to reject it. Carl Gauss knew this, but he was not the first person to try to understand what it means to reject the postulate. This was an approach that had been worked out decades before he was even born.

An obscure Italian priest named Giovanni Girolamo Saccheri started with the same idea John Wallis had in the seventeenth century—to prove the fifth postulate by looking at the possibility of similar triangles. The basis of Wallis's work was his statement "for every figure there exists a similar figure of arbitrary magnitude." Saccheri wanted to start there, but he went a lot further than Wallis. What Saccheri realized was that the existence of similar triangles was itself something that needed to be proven—a postulate. But what if there were no similar triangles? What if the sum of the three angles of a triangle changed depending on the size of the triangle?

This sort of thinking inspired Saccheri to attempt the old mathematical trick of *reductio ad absurdum*—proving something by examining what it would mean if it were not true. If the fifth postulate was not true, Saccheri reasoned, then the geometry that resulted was of a new type, a non-Euclidean form. He called it "universal geometry," and he reasoned that if he could examine this new geometry and demonstrate its absurdity, it would show the necessity of the fifth postulate. Saccheri published a book titled *Euclides ab Omni Naevo Vindicatus* (Euclid Freed from All Flaws) in 1733.

Saccheri was someone for whom breaking the fifth postulate was not just a matter of thinking up some new mathematics. There was much more at stake for him. He struggled with what he was thinking about—he may have even been afraid of it. As a Jesuit, he was bound to answer to a higher authority, whether God or man. Failure was not just an academic result. It bordered on heresy.

Because Saccheri was a Catholic priest, he had to obtain the so-called *Superiorum permissu,* or papal permission to publish. This was no easy matter. The process started at the local level and worked its way up the church hierarchy. First Don Gaspar, a doctor of law and sacred theology, would examine the work to ensure that it contained nothing contrary to the orthodox faith. Next Ignatius Vicecomes, provincial of the Jesuits, had to approve of the grant of permission. Then the Jesuit senate had a chance to weigh in. Following this, the regional cardinal had to sign off on the book. And finally, the inquisitor general, Sylvester Martini, needed to approve it.

If they only knew what Saccheri had written, all those priests and members of the church hierarchy might have condemned him or worse. The Italian philosopher Giordano Bruno was burned at the stake saying things that were less bold and daring than what Saccheri implied in his "universal geometry." He brought the reader to the very edge of questioning the true nature of space and Euclidean geometry. The mathematician Augustus De Morgan once wrote that it would be as easy to rewrite the New Testament as it would be to change Euclidean geometry in any way, and that is exactly where Saccheri brought the reader. But he held back and never plunged over the edge.

Perhaps what comforted the church officials who reviewed Saccheri's work was that it ultimately made the case for Euclidean geometry. After seventy pages of developing his new type of geometry, Saccheri declared it to be false. Why? He had worked out what this universal geometry would be like, and he found it to be absurd. The normal, rectilinear shape of space that we know from our world would be curved, and in this odd system parallel lines would not need to be straight. Because of this absurdity, Saccheri felt confident enough that he had provided proof of the fifth postulate.

Such a declaration has been seen as an afterthought by some—perhaps window dressing to assure its passage by the church authorities who reviewed it. Some have even wondered with hope whether Saccheri actually had some sense of what he had done. Perhaps he understood the nature of non-Euclidean geometry and knew it was a wonderful, new, and completely logical geometrical system. Perhaps he didn't. Whether hiding behind his fear or unable to overcome the limitations of his own imagination, he probably never got past the idea that Euclid was right. What Saccheri never realized was that he had inadvertently stumbled upon something more sublime than absurd—and nearly a century before his time. He invented non-Euclidean geometry by accident, but the strange thing is that he didn't even know it.

Saccheri wrote his masterpiece as a final act of his life—almost an epilogue. Within a year of its publication he died, and his work was all but forgotten for more than 150 years. He was not published for the first time until 1894—and then in an American magazine. For the next several decades after him, many more mathematicians continued in vain to try to prove the fifth postulate. "Not a year goes by," one wrote in the early nineteenth century, "without a book being published on the parallel postulate."

Even if not many people got their hands on Saccheri's work, a few important German mathematicians did. One of them was a professor at the University of Göttingen named Abraham Kästner, who spent years collecting information on the subject of the fifth postulate. All of his research convinced Kästner of one thing—that trying to prove the postulate was pointless. He probably didn't doubt Euclid—he just recognized that enough brilliant mathematicians had worked on the problem through the years, and this convinced him that it couldn't be done.

This didn't stop people from trying. A student of Kästner's named Georg Simon Klügel wrote a dissertation in 1763 called *Conatuum Praecipuorum Theoriam Parallelarum Demonstrandi Recensio* (Review of the Most Celebrated Attempts at Demonstrating the Theory of Parallels). It detailed thirty previous separate attempts to prove the fifth postulate. Saccheri's book was one of the many attempts at proving the postulate that Klügel studied in his thesis. Klügel carefully analyzed Saccheri's work and concluded that Saccheri's proof was no more valid than any of the others. The conclusion he seems to have arrived at was to question whether the fifth postulate could be proven at all. Klügel firmly placed the fifth postulate in the realm of the senses. Before him nobody had ever doubted—at least in print—that the postulate could be proven.

Klügel's dissertation came into the hands of John Henry Lambert, who studied it and praised it in his *Theory of Parallels*, which was published in 1766. "The proofs of the Euclidean postulate," wrote Lambert, "can be developed to such an extent that apparently a mere trifle remains. But a careful analysis shows that in this seeming trifle lies the crux of the matter; usually it contains either the proposition that is being proved or a postulate equivalent to it."

Lambert laid out the essential problem by asking if the fifth postulate can be derived at all, and then asked a question along the same lines as Saccheri—what would you get if you throw it out? This question led him to a strange form of geometry that was really a precursor to non-Euclidean geometry. He anticipated many of the results of non-Euclidean geometry, and some even consider him the first person to have discovered the solution to the fifth postulate.

Still, Lambert didn't seem to know what to make of his own work. Some of the perplexing conclusions of non-Euclidean geometry seemed preposterous to him. This confirmed the ridiculousness of it for him.

Independent of anyone else, the British mathematician Thomas Reid came up with his own version of non-Euclidean geometry about thirty years later, calling it "the geometry of the visibles." He quietly published his work in 1764, and it was duly forgotten as well. Reid still did not believe it. He thought that Euclidean geometry was reality and what he had discovered was some sort of alternative and bizarre geometry.

One of the few places where the work of Saccheri and the others was known was at the University of Göttingen where Gauss was a student. Gauss may have known of Saccheri's work and may have even studied it because Saccheri's obscure book was in the mathematics collection in the library, and Gauss certainly owned another book that referred to the work of the Italian priest. In fact, library records show that Gauss found at least one other book about the fifth postulate in the university library and checked it out twice while he was a student.

In the second decade of the nineteenth century, Gauss began to record and discuss his sense of the futility of continuing to try to prove the fifth postulate. Taking him at face value is difficult because he never drew his ideas together into a single cogent argument. All we have are scattered notes, inferences, and occasional remarks to friends in letters. In 1813, he wrote in his notes, "We are now no further than Euclid was. This is the *partie honteuse* [shameful part] of mathematics, which sooner or later must get a very different form."

A few years later, in 1816, Gauss showed that he had even darker thoughts about geometry. He wrote to his student Christian Gerling that he thought there was nothing absurd in denying the fifth postulate. A year later, he wrote to Olbers, "I am becoming more and more convinced that the necessity of our geometry cannot be proved." In the same letter he said, "Perhaps only in another life will we attain another insight into the nature of space, which is unattainable to us now."

Still, with the exception of these and a few other letters, Gauss declined to publish anything at all on the matter—for reasons that historians are at a loss to explain. He was satisfied to leave it be, avoid controversy, and go on with other work.

Perhaps Gauss was so prolific in so many other areas of mathematics that he simply may not have had time to finish his work in geometry. Other universities were still trying to lure him away, even after he had settled at Göttingen. Or it may be that he was not satisfied enough with his material to let it out into the world. Plus he now had the added responsibility of raising a family and attending to the administrative duties of being a professor. Like one of his idols, Isaac Newton, Gauss was not very fond of the administrative day-to-day.

Gauss idolized Newton because Newton, like Gauss, was an agile genius whose mathematical mind was unrivaled in his day. And like Newton, Gauss was a great believer in perfection in his work. He sought it—needed it—in what he wrote. He idolized Newton for his celebrated works, which were products of years of hard work and effort.

Unlike Newton, whose genius was not truly celebrated until he was middle-aged, Gauss was barely into his twenties when his scientific celebrity rippled throughout Europe. This may have had a profound effect on him by setting the bar even higher. He was driven by a complete love of perfection—perhaps even hampered by his need to attain it.

Possibly there was another reason for Gauss's reluctance to publish anything on non-Euclidean geometry. He may have been reluctant to go against two thousand years of established wisdom—indeed, against reality itself. This was a reality to which, by all indications, Gauss himself subscribed. He saw mathematics as a basic method for understanding nature. Non-Euclidean geometry flew in the face of that, and it flew in the face of the most influential German philosopher of Gauss's day—Immanuel Kant.

Kant's latest work, the *Critique of Practical Reason*, was still just a few years off the presses when Gauss started college. It followed his monumental *Critique of Pure Reason*, which had a profound effect in intellectual circles throughout Europe, especially in Germany. Göttingen was no exception. There, Kant was king. He asserted in his *Critique of Pure Reason* that the universe is Euclidean. Gauss read the *Critique* five times because he was interested in how Kant conceptualized space. Kant said that Euclid was right. Who was Gauss to say otherwise?

He may even have been afraid of the uncertainty the controversy would stir. Certainly some of the commentators who tried to make sense of this after he died put it down to cowardice on his part. Why would he discover something as novel as non-Euclidean geometry and not tell anyone anything about it? Gauss himself said that he feared the "clamor of the Boetians," a reference to the ancient people of Boetia, who are said to have been dull-witted and boorish.

Maybe this was not so much fear as it was the desire to continue his work in peace. This clamor, had it erupted, would surely have driven Gauss to distraction. He was a great mathematician—and a historic figure—but in the end he was a simple man given to the simple taste of hard work. He applied his vast theoretical mathematical skills to some of the greatest problems of his day, and he was happy to do his work in peace. Little did he know that thousands of miles to the east, a talented mathematician at a remote Russian university was about to disturb the peaceful Euclidean universe.

9

Lessons of Curvature

Only impractical dreamers spent two thousand years wondering about proving Euclid's parallel postulate, and if they hadn't done so, there would be no spaceships exploring the galaxy today.

—Marvin Jay Greenberg

When Czar Alexander I came to claim the Russian throne in 1801, one of his first ideas was to develop and modernize the educational system in his dramatically undereducated country. This meant establishing new universities in several cities, including one in 1804 in the city of Kazan—the ancient capital of the Tatar kingdom in central Russia, about three hundred miles east of Moscow.

Alexander modeled Kazan on the German system of universities. German institutions were among the best in the world at the time, and hoping to emulate that success, University of Kazan professors would be expected, like their counterparts in Germany, to conduct research along with their teaching duties. In some cases, the new Russian professors would be German, with Kazan recruiting several to fill its ranks.

As a professor teaching in Leipzig at the time said, "The teaching of physics and mathematics in Kazan was as good as at any German University." There was a professor named F. K. Bonner, who taught physics; I. A. Littrov, who taught astronomy; and K. F. Renner, a professor of applied mathematics. And there was Johann Christian Martin Bartels, who arrived at the University of Kazan in January 1808.

This was the same Bartels who a couple decades earlier had been an assistant to Carl Gauss's first teacher, Mr. Büttner, and had been assigned a tutor to Gauss. Living close by and being very interested in mathematics himself, Bartels had a profound influence on young Gauss, bringing him into contact with the mathematics to which Gauss would devote his life. Gauss had a profound influence on Bartels as well, for after this first teaching experience, Bartels began a lifelong love affair with mathematics. Mainly a teacher of mathematics, Bartels would be one of the most influential mathematicians of his day through his students.

After his mentorship of Gauss ended, Bartels made his way to Switzerland, spending nine years there before returning to Brunswick just in time to welcome Napoleon's troops. He spent a few years there but left Germany in 1807 to migrate east, having been lured by the diamond of Czar Alexander's newly created University of Kazan. He accepted a post as professor of mathematics there, and he soon met Nikolai Lobachevsky, becoming his teacher.

Lobachevsky was born in 1792, shortly after Bartels finished his duties with Gauss and just before Gauss started college. Coincidentally, this may have been also about the same time that Gauss first conceptualized what it would mean if the fifth postulate were not true—the first step toward developing non-Euclidean geometry. Lobachevsky would eventually come to traverse the same path and go even farther than Gauss. His journey was not easy.

Lobachevsky's father died when he was seven, leaving his mother a widow raising three boys. Lobachevsky's boyhood at that point suddenly shifted from poverty to extreme poverty. Fortunately for him, he did well enough in his early studies that he was able to enter the university as a charity student in 1807. The university had a tiny class initially. In 1809, there were only forty students enrolled.

At Kazan, Lobachevsky was initially interested in studying medicine, but that all changed when he came under the influence of a few foreign professors, including Bartels. In this, Bartels had the unusual fate of teaching the chosen art to two of the greatest mathematicians of his day—first Gauss and now Lobachevsky. This would be by far the most important role Bartels would play in the development of mathematics in the nineteenth century. He would also profoundly influence both men and in doing so affect the development of non-Euclidean geometry.

Bartels was the best teacher Lobachevsky and his classmates could have had. He introduced them to the works of Newton, Euler, and Lagrange and more than a century and a half of mathematical analysis. He also brought Gauss's work with him from Brunswick, and introduced it to Lobachevsky and his classmates. Lobachevsky quit medicine and threw himself completely into mathematics.

As Lobachevsky rose through the ranks of mathematicians at Kazan University under the guidance of Bartels, young János Bolyai was being homeschooled by his father during the Hapsburg Empire, in what is now Romania.

Bolyai's mother was said to have been wonderfully beautiful, fascinating, and incredibly smart. His father, Farkas, was one of the great mathematicians of his age. Noble and brilliant, János was the perfect combination of the two of them.

The ancestors of the family had once lived fabulously, inhabiting a castle in the town of Bolya. Sometime more than a century before, however, one of the progenitors had been imprisoned in Turkey and lost the castle, and the family began a slow decline into poverty. By the time János was born, the Bolyais had little money. Still, Farkas had a good job. His possessions included a house, a garden, some money, and a stipend of pork, lamb, salt, honey, wine, flour, and firewood. Still, Farkas and his wife were unhappy. She had mental problems and her mother interfered in the marriage.

János was homeschooled until he was nine years old, but not in mathematics. It was not until he was nine that his father, Farkas, began to teach him mathematics. This seems a strange decision given that Farkas was a teacher of mathematics, physics, and chemistry and perhaps one of the most knowledgeable and gifted mathematicians of his day. His son showed remarkable ability in all manner of subjects, gaining the ability to read by the age of five and playing violin well by the time he was seven years old. Why not teach him mathematics as well at an early age?

Perhaps Farkas waited until his son was nine because he didn't want him to catch fire too quickly and burn out. In any case, despite his late start, young János mastered geometry and calculus within a few years. "He works in them with extraordinary readiness and ease," Farkas beamed in a letter when his son was only thirteen. "He has a quick and comprehensive head, and often flashes of genius, which many paths at once with a glance find and penetrate."

Farkas was not just spewing empty boasts. That same year, he was unable to make it to one of his classes, and he sent his son in his place. Most classrooms of college students would probably bristle at being taught an advanced lesson by a thirteen year old, but young János did so well that some of Farkas's students said they actually preferred him to his father. These talents extended much further than the classroom. As the boy grew into a young man, a soldier, he would be dashing, skilled, and smart. He would foster a passion, like his father, to prove the fifth postulate.

János Bolyai grew up with a picture of Gauss on his wall. These two influences—the real force of his father and the virtual influence of Gauss's ever-present image on the wall—would almost guarantee that János would grow into a mathematician himself. He would later be called the greatest Hungarian mathematician of all time and the man who changed space.

His story is one of hardship and disappointment despite his success, though. Shortly after he turned twenty, János would invent non-Euclidean geometry in his spare time while serving as a junior army officer. In his early thirties he retired on a pension as a semi-invalid, slowly surrendering his happiness to a lifetime of disappointment.

Gauss hemorrhaged his domestic bliss all at once. His life took a tragic turn when his wife died on October 11, 1809. A month and a day before, she had given birth to their third child, a son, and the strain of this birth proved to be more than Johanna could take. She died, and the weak child—"poor Louis" as Gauss called him—followed her a few months later. Gauss was left a widower raising his two little children while adjusting to a new job in a new city.

His thoughts were not on his own plight except inasmuch as it involved losing Johanna. "I closed her angelic eyes in which I have found heaven for these last five years," he wrote to his friend Olbers the day after she died. "Heaven give me the strength to bear this blow." His prayers went unanswered. After Gauss died and his papers were collected and studied, a few long journal-entry-like letters were among them—some of which appear to be blurred by tears. In one of these passages he wrote, "Even the serene sky makes me sad." Sometime thereafter Gauss wrote a tribute to his wife, which was not discovered among his papers until long after he died.

In the passage he extols her virtues and laments his loss. "How could I have deserved you?" he wonders.

Gauss did not stay a widower for long. He was famous enough that after his first wife died, he soon remarried. The bride was Minna Waldeck, the daughter of Gauss's older colleague Johann Peter Waldeck, a professor of law at Göttingen. Minna had large, probing eyes and a big, pretty nose. She had been friends with Johanna Gauss and had been previously engaged to another man. She was coming off a bad breakup of a relationship that had started in courtship, progressed into engagement, and ended on the rocks in the weeks before Gauss proposed. This bad ending had left her depressed, and it was only through the intercession of her parents that she agreed to marry Gauss.

Theirs was a very sudden and strange pairing. She was not of the same social upbringing. Her parents were more well-to-do and much more educated than Gauss's parents. This came out in embarrassing ways sometimes, such as when early in their courtship Minna wanted to write to Gauss's mother. Gauss had to ask her not to, apologizing that his mother was illiterate.

Despite the whirlwind speed of their courtship and the differences in age and upbringing, Gauss certainly had the approval of Minna's father. He was, after all, one of the most famous professors in Germany by now and had a reputation for unassailable scholarship and long-standing favor in the royal courts. Gauss was motivated by the need to have a new mother for his children. He was stunningly honest about this fact in the letters of courtship that he sent to Minna. "I honor you too much to wish to conceal that I can offer you only a divided heart from which the image of the glorious shadow [of Johanna] will never disappear." Minna would have to live under that glorious shadow as well, and she would spend the next few decades obscured by it. On August 4, 1810, less than a year after Gauss's first wife died due to complications in childbirth, the two were married. That same year, Gauss was elected to the prestigious Institute of France. He and Minna started having children right away. Their son Eugene was born the year after they married. Two years after that, another son was born. In 1816, Gauss's wife gave birth to a daughter, Theresa, their youngest child.

Gauss started the second decade of the nineteenth century with a new lease on life. He still had a relatively new job and now a new wife and more children added to his family. He was spearheading an ambitious effort to build the new observatory at Göttingen, under the most difficult of circumstances. He started out under the regime of the Westphalian government, and his observatory might have never gotten off the ground had someone else been at its helm. Gauss managed to garner significant funding for the observatory during the next several years. Even after the fall of the occupational government in 1814, he continued to buy equipment for it.

Lobachevsky, meanwhile, was in trouble for much of the same decade. He was a rabble-rouser. He was arrested and imprisoned while a student for setting off a massive firework over the town of Kazan. As punishment for his many slights, he was often reprimanded and forbidden attendance at certain events. When Lobachevsky was eligible for his master's degree in 1811, things came to a head. He crashed a masked ball around Christmastime and got in trouble for this as well as for supposed atheistic statements involving God's role in the weather. The university senate refused to grant him the degree. He was almost expelled from school altogether.

An assistant inspector named P. S. Kondyrev reported that Lobachevsky was showing signs of atheism. This was a dangerous report because there had been an official decree by Czar Alexander I in 1811 that university officials were to expel students who showed any such tendencies and immediately draft them into the army. This was the time of the Napoleonic Wars, so serving in the Russian army was a terrible fate.

Lobachevsky's commandant complained to the rector. The rector complained to the curator. On August 23, 1811, the curator presented a complaint to the university council about Lobachevsky, calling him "the leader in bad conduct." They added that they regretted how badly this conduct clouded his intellectual abilities. The complaint eventually reached the minister of education, and Lobachevsky might have lost his place in the university had it not been for the mathematics professors, who banded together and appeared before the senate on his behalf. In the end, he was saved by the teachers whom he would later eclipse. "I am delighted by the success of my work," wrote Bartels about their work together.

In the end, Lobachevsky avoided conscription into the Russian armed forces and graduated after he promised to behave himself and swore off atheism. From there, he rose through the ranks very quickly. He earned his master's degree in 1811, became a lecturer in 1814, and was a professor two years later. This job consumed him almost completely, but he made his mark and distinguished himself as a great teacher. He taught civil servants, who according to a recently passed law were required to have a college degree. At a time when most professors taught directly out of approved textbooks written in Latin, Lobachevsky wrote his own lecture notes and delivered them in Russian.

Lobachevsky's lectures on geometry helped him rise through the ranks to become one of the top academics in Kazan. He became a professor in 1816 at the age of twenty-three. That he wrote in Russian must have helped, because after 1815 the official language of his country's universities became, by decree, Russian. And ironically, after coming to Lobachevsky's rescue just a few years before, the foreign professors were expelled from Kazan in a purge to eliminate all non-Russian influences. Lobachevsky could do nothing but stand by and watch.

In the midst of all this, around 1815 Lobachevsky became interested in the fifth postulate as part of his work on geometry. Initially, like everyone else, he was interested in proving it, and he would spend the next ten years trying to do just that. As he was just finding his way toward the questions in geometry that would forever define him, Gauss was already there.

On the surface, Gauss's main interests in these years were practical and experimental. This was an incredibly busy time for him professionally and personally. He was overseeing the construction of Göttingen's new observatory, which was built just outside the city gates in 1814. Gauss and his family didn't move in until the fall of 1816. By that time he and his second wife had three children in addition to the two older children left from his first marriage with Johanna. They would move into the observatory and raise them there.

At the observatory, Gauss spent a great deal of time and energy engaged in astronomy and the improvement of astronomical instrumentation. But he didn't neglect the theoretical work. He worked heavily on topics in pure mathematics, including on non-Euclidean geometry.

His letters from this period reveal how much his thinking had evolved with regard to geometry. A decade earlier, around 1808, Gauss had come to fully appreciate the utter futility of trying to prove the fifth postulate, and by 1813 he had taken to calling it the "shameful part" of mathematics.

By 1816, Gauss was on the verge of a revolution in mathematics, but he was no mathematical revolutionary. This was a revolution that he would ignore, at least for the next few decades. He saw mathematics as the backbone of physics—once suggesting that all true physicists had to be thoroughly grounded in mathematics. But the revolution that Gauss was realizing in 1816 suggested the opposite as far as geometry was concerned. The revolution of non-Euclidean geometry asked the question, Does mathematics need be grounded in physical reality?

Non-Euclidean geometry was a strange revolution, because it went against the practical nature of mathematics and denied the physical reality of geometry itself. It required one to forget what one had been taught and what was presumed about geometry—about the world.

The basis of applied mathematics is to represent a real-world problem in mathematical terms and then find a real-world solution by solving the problem. Geometry was the original applied mathematical system when it evolved under the Egyptians, where it was literally a set of tools adapted for measuring land. Under the Greeks it grew to be a rich set of theories underlying the applied aspects, but it never really shook itself free from application. In a sense, applications legitimized geometry. The field evolved before any sense of scientific theory existed, and one result of this was that nobody would question whether geometry was true or not.

Realizing this by the end of the 1810s, Gauss could have shaken the mathematical world by publishing his ideas. But he published nothing—another thing that seems strange today when the pressure to do so is much greater. Today it's not uncommon to rush publications into print, or online, sometimes even before peer review is finished and corrected proofs are available.

Gauss was, in general, reluctant to ever publish anything. Some of the greatest mathematicians of his day were excited to make great discoveries only to find that Gauss had made the same conclusions but didn't even bother to publish them. His friends were aware of this self-stifling tendency and deeply regretted it because they knew how brilliant he was and wondered what sort of treasures might lie unexposed in his notebooks. Gauss was fully aware of this tendency in himself as well—and unapologetic besides. Once, when he admitted to his friend Heinrich Christian Schumacher his reluctance to publish, his friend wrote to him that he wished, for the good of science, he were not so reticent. "We would then have more of the infinite richness of your ideas than now," Schumacher wrote. "And to me the subject matter seems more important than the most complete form of which the matter is capable."

For Gauss, there would be no publishing of ideas that were not fully formed. On another occasion, he explained this in a letter to his friend and pupil Johann Franz Encke. "I can have no real joy in the incomplete," he said, "and a work in which I have no joy is a torture to me." For Gauss, a complete work was nothing short of groundbreaking and brilliant. His thoughts on the fifth postulate were not mature enough to make him comfortable publishing anything in 1816. Still, one wonders if he would have published his thoughts on the fifth postulate even if he were inclined to do so. Because of the strange nature of the subject, this was perhaps the most controversial work he would ever engage in.

He was no more comfortable sharing his work privately with others. He had endured a painful episode at the turn of the nineteenth century when the adviser Eberhard von Zimmermann, who worked for his patron Duke Ferdinand, gave some papers with Gauss's work on the seventeen-sided figure to a Prussian army officer. This officer took this work and published it. "This person," Gauss later complained, "had the impudence afterwards to publish a work, which I have not myself seen, where he sets forth this matter in such a way that the reader might well believe that it was [his] own work."

In fact, Gauss was wrong about what had happened. Had he actually seen the publication he would have read a full acknowledgment from the Prussian military officer. "The regular 17-sided polygon . . . was first discovered by a young geometer of Brunswick whose name, as I remember, is Gauss," the officer wrote.

Still, the damage was done. Gauss had experienced that feeling of being violated that comes from being a victim of plagiarism. In any case, even if Gauss were completely comfortable with his ideas, he was entering the busiest time of his professional career and personal life. He had five children at home and an observatory to run. Gauss once complained to the mathematician Friedrich Bessel that he was in need of more time than he could imagine. "And my time is often limited, very limited," he added. In the end, he never did publish anything, instead writing cryptic notes to a few friends and leaving scattered references to geometry among his papers.

In 1816, Gauss's old friend Farkas Bolyai was about to reenter his life. Farkas was one of the few people with whom he had ever shared his intimate thoughts on the subject, and though the two had lost touch for nearly a decade—years that encompassed the rise and fall of Napoleon—they had been following parallel tracks in thinking about geometry. During this time, Farkas had spent untold grueling hours trying to prove the fifth postulate, and his efforts had all ended in frustrating failure.

Farkas was too smart to fall into the self-deluding trap of believing he had solved the fifth postulate as had so many before. Numerous mathematicians, after years of work, had convinced themselves that they had found the answer, only to be proven wrong later. Farkas was not so charmed by his own brilliance to be fooled so easily by his own work. This was his depressing lot. He was not smart enough to solve the fifth postulate but too smart to convince himself that he had. In 1816, he knew that much of the work he did in the previous two decades had been a waste of time.

Looking back on these wasted efforts must have been hard for Farkas to take because he was a proud, dignified, accomplished man. He was a talented playwright and still holds a place of honor in the annals of Hungarian literature. He was also an inventor and created contraptions such as a large horse-drawn sleeper carriage—perhaps the world's first mobile home.

Farkas admired Gauss for not having spent so much time on the problem of the fifth postulate as he had in the previous decade. Gauss never would have been able to finish the important books he penned if he had wasted the time on the postulate, Farkas once wrote. But he had no idea that Gauss had in fact been thinking about the fifth postulate as well, working on and off for much of the previous decade. Had Farkas known, he might have written Gauss sooner.

In 1816, when Farkas wrote Gauss his first letter in a decade or so, he did not even mention working on the fifth postulate. He was concerned with a more pressing and practical question about what to do with János, his teenage son. Farkas and his wife didn't have enough money to send their son abroad for his education. This presented a problem because the local universities did not have strong mathematics traditions. Farkas was reluctant to send young János to any of them.

So Farkas did what any father would do in the situation and sought help by reaching out to a friend—in this case, Gauss. He wrote to Gauss in the spring of 1816 proposing that his fourteen-year-old son travel to Göttingen and study under the famous mathematician. Perhaps Farkas would have been more successful if he had included some mention of his own mathematical work in the previous decade—particularly his continued efforts to prove the fifth postulate. This would have been something near and dear to Gauss's heart.

In fact, that same year, Gauss wrote book reviews of two separate works whose authors had attempted to prove the fifth postulate. He was completely convinced that the postulate was impossible to prove and that any effort to do so was doomed to failure. He decried what he called any "vain effort to conceal with an untenable tissue of pseudo proofs the gap which one cannot [fill]." But Gauss never got a chance to discuss any of this with his old friend. He didn't even answer Farkas's letter.

The problem may have been that Farkas seems to have fallen into some sort of dark misogynistic space when he was writing the letter. It was filled with bizarre passages, perhaps informed by his problems with his wife. In the letter, he asked Gauss repeatedly about Minna. Was Gauss's second wife exceptional among women? Not more changeable than a weathervane? Not unpredictable like a barometer?

Gauss ignored him, and what might have been the greatest teacher-student team in history never came to be. And perhaps one must pity young János Bolyai. He grew up with a picture of Gauss on his wall, but he never had the opportunity to learn caution from him. Gauss was a mathematical matador, János a bull running wild through the streets of Europe.

10

To Stir the Nests of Wasps

I hope that posterity will judge me kindly, not only as to the things which I have explained, but also as to those things which I have intentionally omitted, so as to leave to others the pleasure of discovery.

—René Descartes

Carl Gauss's old, blind mother, Dorothea, moved in with him in 1817. She was a simple woman—a hard, strong woman—nearly bent in half by the humility of a harsh German peasant existence. And to Minna and the children, she became a permanent guest, living with them at the Göttingen observatory for the remainder of her long life.

Every morning and every night there Dorothea was, dressed in the sad old peasant attire of two generations before, living a simple life on the periphery of the household. Strangely, she didn't take her meals with the rest of the family. Minna must have heard her scuffling around the cold observatory a thousand times, feeling her way along the walls in her blindness. This went far beyond the normal mother/daughter-in-law dynamic. She really didn't have much in common with the old lady.

The one thing the women did have in common was Gauss himself. He was the observatory director, scientist, father, husband—but perhaps also somehow still the little boy who had long ago left his mother's home. He had never been particularly close to his parents. His brilliance shone early and illuminated education opportunities that pulled him increasingly far from his parents and their poverty. He departed their economic station altogether.

Dorothea was a devoted mother and was especially proud of Gauss's academic achievements. She had been a maid for seven years before marrying Gauss' father—it was an unhappy marriage— and she probably would have lived her days in anonymity and misery had she not given birth to one of the most successful mathematicians in history. Even before Gauss's genius emerged, Dorothea adored her only son completely. And Gauss was the sort of boy a mother couldn't even dream of. Once when he was home from university with a good friend, Dorothea asked his friend whether Gauss would ever amount to anything. The friend told her that Gauss would become the greatest mathematician in Europe, and Dorothea immediately burst into tears.

She was central in Gauss's life, and in her later days, blind and unable to live on her own, she became even more so for her dependence on him.

These were the dying days of the fifth postulate. The mystery was about to be solved; everything in mathematics was about to change. With the exception of Gauss and a few others, nobody had any idea what sort of storm could be brewing in the field. By the time Dorothea moved in with Gauss, Gauss had probably convinced himself that the fifth postulate could not be proven and that a new, non-Euclidean geometry might exist. His methods are a little obscure because he never published anything, and the only notes that survive are ones that he wrote down decades later, but he had worked intermittently on non-Euclidean geometry from 1813 to 1816. He had tried to make geometric constructions to deduce the fifth postulate. He managed to convince himself that the postulate could not be proven and was not necessary at all.

Still, he prevented himself from stepping forward and publicly pronouncing the fifth postulate dead. Gauss was also well aware of the controversy of saying anything about non-Euclidean geometry. "I am glad that you have the courage to express yourself as though you recognized the possibility that our theory of parallels, consequently all our geometry, might be false," he wrote to his friend Christian Gerling in 1818. "But the wasps, whose nests you stir up, will fly around your head." Gauss was not about to stir up these wasps himself—at least not directly.

He would, however, comment on the work of others as the opportunity arose. Having decided that the fifth postulate could not be proven, woe unto those who tried to do so! Gauss was known to be a severe critic of others—to the point of unfairness at times. He sometimes gave harsh verbal assessments of his contemporaries. But the same was not necessarily true of his written reviews. Gauss was an active reviewer of the mathematical publications of his day, and in these he usually abstained from negativity. As a writer he was quite gifted, and the language of his first book was most refined. From there he set the standard as a scientific writer, influencing several generations of scientists to come to Germany. He avoided the use of repetition or any rhetorical devices in his works, and of the work of others he was generous when impressed with it. Then he would often write in language full of praise.

In 1816 Gauss had several opportunities to turn these gifts toward the efforts of others who were attempting to solve the fifth postulate. It was a banner year for false proofs, and two more attempts appeared in Germany. Gauss reviewed these works, books by the mathematicians J. C. Schwab and Count Metternich, and in his reviews he suggested that the fifth postulate couldn't be proven. This stirred up a little controversy, and some criticized Gauss for thinking it.

Gauss also intimated his belief in the possibility of a new, non-Euclidean geometry. He had the intuition that if the fifth postulate were to fall, a new non-Euclidean geometry would follow, but this was an idea too radical for him to accept. So his testing of the waters of publishing was a failure, and he would decide to never again publish anything on the subject in his lifetime and keep his secret obsession secret. For once, he was not alone. One of his former students had just reviewed the same works and come to a similar conclusion.

Friedrich Ludwig Wachter was a student of Gauss's in 1809 at Göttingen. He was a smart young man at a difficult time in German history, and his career had been sidetracked by the Napoleonic Wars. Gauss had helped him find a professorship, but he had to leave his post in 1813 for army service.

Having completed his tour of duty, Wachter was appointed a professorship in Danzig soon thereafter. He came to visit Gauss in

1816, and having just reviewed one of the works Gauss had reviewed, the two discussed the fifth postulate. Wachter was beginning to think about something he called anti-Euclidean geometry, and he wrote to Gauss toward the end of 1816 spelling out his ideas. He thought he could prove the fifth postulate indirectly, by showing that it had to be true.

In this letter, Wachter speculated about a form of geometry on the surface of a sphere. As the sphere got larger and larger, the surface would appear more and more flat. The theoretical end of this would be when the sphere became infinite in radius. This infinite sphere meant that geometry as it was known could be true even if the fifth postulate were false because the curvature would be zero in an infinite sphere. Even if the fifth postulate could not be proven true, drawing parallel lines on the surface of the infinite sphere would guarantee that it would hold. "The Euclidean geometry is false," Wachter wrote. "But nevertheless the true geometry must begin with the same [fifth postulate] or with the assumption of lines and surfaces which have the property presumed in that axiom."

Gauss wrote a few months later that Wachter's proof was incorrect. He recognized the problem right away. "Though Wachter has penetrated farther into the essence of the matter than his predecessors, yet is his proof not more valid than all others." This turned out to be generous. A century later, a historian of mathematics called Wachter's attempt to justify geometry a "monstrous conglomerate blunder."

Nevertheless, Gauss was encouraged by Wachter's work and the possibility that there was more to geometry than Euclid had realized. He wrote to his friend Heinrich Olbers in 1817, "I am coming more and more to the conviction that the necessity of our geometry cannot be proved, at least not by human intelligence nor for human intelligence. Perhaps we shall arrive in another existence at other insights into the essence of space, which are now unattainable to us. Until then one would have to rank geometry not with arithmetic, which stands a priori, but approximately with mechanics."

For his part, Wachter was encouraged by Gauss and began to consider his "anti-Euclidean" geometry. He didn't get very far, though. He disappeared without a trace the next year, the victim of some unknown accident or crime. His whereabouts remained a mystery, and after about ten years he was finally officially declared dead.

●　●　●

The next person who came close to solving the fifth postulate was a man named Ferdinand Karl Schweikart. At around the same time Wachter was meeting with Gauss, Schweikart was growing suspicious of all previous attempts to prove the fifth postulate. By 1816, he developed what he called "astral" geometry. He wrote a treatment of this astral geometry, essentially spelling out the ideas of non-Euclidean geometry.

Schweikart wrote a paper in 1818 that developed his theories. As the Italian priest Giovanni Saccheri had done nearly a century before, Schweikart began by assuming that the fifth postulate is not true and then examined the resulting effect on geometry. He came up with a new way of looking at the problem by using a parallelogram—a four-sided figure where each pair of opposite angles is equal. Exploring this, he had become convinced that Euclidean geometry is "only a chapter of a more general geometry." He postulated that under certain conditions, Euclidean geometry holds, while under other conditions astral geometry takes over. The idea was that geometry was composed of the hypothetical true geometry and the practical (or Euclidean) geometry. "There is a two-fold geometry—a geometry in the narrower sense," Schweikart wrote in his paper. This narrower geometry he called "the Euclidean and an astral science of magnitude." He chose the name, apparently, to refer to the celestial world.

Schweikart sent his book to Gauss's friend Gerling and asked that he forward it for comment. Gerling wrote Gauss a letter in 1819 alerting him about this new figure on the scene who called his work astral geometry. "I learned last year that my colleague Schweikart formerly occupied himself much with mathematics and particularly has also written on parallels. So I asked him to lend me his book," Gerling wrote to Gauss.

After reading the short article written by Schweikart, Gauss was thrilled. He replied to Gerling that the work was almost as if he had written it himself. By now, Gauss knew for certain that Euclidean geometry was not complete. In reading Schweikart's article, what Gauss had once called the "shameful part" of mathematics now must have seemed all the more shameful. People had spent thousands of years trying to prove something that was not true.

Schweikart had invented his astral geometry sometime in the years between 1812 and 1816, and some would consider him to be the first to write a treatise on non-Euclidean geometry. He certainly was the first person to write a treatise consciously aware of trying to explore a new world of geometry for its own sake—not simply as a way of trying to prove the fifth postulate by reducing the argument to the absurd. His work persisted for several years. In 1829, for instance, the mathematician Friedrich Bessel wrote to Gauss and underlined the need for a correction to geometry, based on the work of Schweikart and others with whom he and Gauss were familiar.

Gauss was very pleased with Schweikart's work. Its one weakness was that Schweikart was a lawyer and not a mathematician. This fact would come back to haunt Gauss, because Schweikart had a nephew named Franz Adolf Taurinus who was also fascinated with this area and studied his uncle's book carefully. Taurinus, however, was a mathematician, and so Schweikart encouraged him to pursue the work.

Taurinus was both gifted and hardworking, and he took up the subject in earnest in 1824. That same year, he wrote two papers on non-Euclidean geometry and got in touch with Gauss, who replied in a letter on November 8, 1824. This letter was one of his most explicit communications on the subject. In it, Gauss partially explained his views of the fifth postulate. The letter is also a significant milestone in the history of geometry because Gauss uses the term ''non-Euclidean geometry'' for the first time. ''The assumption [that the fifth postulate is false] that the sum of the three angles of a triangle is less than 180 degrees leads to a special geometry, quite different from ours, which is absolutely consistent and which I have developed quite satisfactorily for myself, so that I can solve [almost] every problem in it,'' Gauss wrote.

Ironically, as significant a step forward as this was for Gauss, it caused him to take three steps backward—because he misread or misunderstood where Taurinus was coming from. The problem was that Taurinus never really bought into non-Euclidean geometry. So when Gauss encouraged him to continue the work, asking him at the same time not to tell anyone what he, Gauss, thought about the subject, he could not have expected what would ensue.

Taurinus continued to work, but certainly not in the way Gauss intended. He published a book in 1825 basically refuting his uncle

Schweikart's work. He never doubted that Euclidean geometry was the true geometry. In fact, he gave numerous reasons why it had to be. Nevertheless, he still thought that the possibility of having a consistent non-Euclidean geometry was real. That this non-Euclidean geometry was not real was of little consequence. He thought it "might not lack significance in mathematics."

Taurinus investigated what sort of results would arise if one were to develop a geometry without the fifth postulate, and he used this as a way of suggesting that Euclidean geometry had to be true— exactly the same sort of thing that Saccheri had done a century before. So once again a daring mathematician had come to the edge of geometry and, staring out over the uncomfortable chasm, decided to turn back to the comfort of the two-thousand-year legacy of Euclidean geometry. Taurinus published two works on the fifth postulate, and he tried to incite Gauss to comment on them. He even suggested in the introduction to one of these pieces that other mathematicians should seek out Gauss's views.

Gauss was made so paranoid by this that he didn't even write back to acknowledge that he had received the books. Neither did he review them or recommend them to others for review. He did not even mention them at all. This proved to be a grave disappointment to Taurinus. His work soon forgotten, he fell into despair. He was so bitterly disappointed to find that one of his two books, titled *Elementa,* had attracted no attention whatsoever that he burned all his copies of the manuscript.

For his part, Gauss was probably happy never hearing from Taurinus again. He may have come to fully understand non-Euclidean geometry, which his notebooks would later reveal when other mathematicians began going through them decades later after he died. But now more than at any other time, he had nothing to say about the subject—not even a hint.

Besides that, he was busy with other things. His job had suddenly started requiring him to travel, to climb countless hilltops and tower turrets, measuring thousands of triangles across the German landscape.

Anyone who thinks of Gauss as strictly a theoretical scientist is wrong. He was very interested in experimental work and instrumentation, and he was a creative genius at times when it came to designing and executing experimental measurements. Proof of this genius is a device he invented after being inspired one afternoon in 1821 while he was walking with his son Eugene. They were talking as they walked, and Gauss was annoyed suddenly by a flash of light reflected off a distant windowpane. In the next flash, an idea came into his head. He could construct a device to do the exact same thing, and this would help him measure the distance from point to point.

Gauss designed a device called the heliotrope, which allowed angles to be measured over greater distances and even in cloudy weather. His invention was first constructed in Hamburg in 1821, based on sketches that he drew. Basically, it was a simple flat mirror about the size of your hand that could be rotated to reflect light from one place to another. By reflecting sunlight toward the horizon, Gauss could have one of his assistants miles away train a telescope in his direction and spot the light.

This simple device greatly enhanced his ability to accurately map two points relative to each other. Using it, he could record the exact angle of reflection from one place to the next, and using the very long baseline between points the heliotrope allowed, he could easily determine the exact direction and elevation from point to point. Gauss knew that using a long baseline like this was the most precise way to make measurements and the best way to make an accurate map.

This work was important to the government because maps that were as accurate as possible were crucial in a land where the preceding centuries had been marked by one war after another. In the post-Napoleonic era, there was a lot of interest in getting the most accurate maps possible, since one of the hardest lessons learned by generals and political rulers everywhere in Europe during the years of the Napoleonic Wars was that maps could be as crucial a military asset as cavalry. After the wars, many of the surrounding governments were commissioning this sort of proprietary map work.

Gauss's interest in making such measurements dated back to the years even before Napoleon's defeat. Not long after his rise to fame after the rediscovery of Ceres, Gauss began experimenting with a sextant that he borrowed from Baron von Zach. He made measurements in the surrounding lands, mapping the lands and dreaming of doing an even larger survey.

His chance came when he was roped into working on a survey with his friend Schumacher, who had been hired by the king of Denmark to accurately map the lands extending beyond Copenhagen. Schumacher suggested that there might be an opportunity to extend this survey even farther and into Germany. This was where Gauss would come in. Already there was a survey of the Bavarian lands, and the Netherlands was also complete. If these two lands could be joined, Schumacher and Gauss would have mapped an enormous area. The idea was to connect and extend these earlier measurements and provide a thorough mapping of the region. Gauss was an enthusiastic advocate of this idea and wrote to his friend as early as the summer of 1816 to say he would participate. In 1818 he became director of a project to map the German lands around him, and he threw himself into the work. So successful was the survey that the king ordered its extension throughout the whole of the province. Gauss was to direct this as well.

Denmark's King Frederick VI had his London envoy approach the king of England (whose dominion included Göttingen) and convince him to fund the survey work under Gauss's direction. The king approved this activity in the spring of 1820, and Gauss began working on it soon thereafter.

It was the most accurate survey of German lands ever attempted. The way measurements were made in those days was by selecting particular reference points, such as the massive stone tower of St. Michael's church in the city of Lüneburg. From a given point at a given latitude and longitude, other points could be mapped by accurately measuring the distance and angles between them, marking out a series of intricately connected triangles among various landmarks. This labor-intensive work was simple in concept, but it took more time than it might seem. The Danish survey was delayed for so long that decades later it was still not finished.

The measurements Gauss was hired to take were just as demanding. During this next decade, he is said to have slept no more than a few hours a night, and he once claimed that he worked harder in these days on this subject than he had ever worked on anything in his life.

Gauss proved more than worthy of the task, but the work took a lot out of him. He made so many measurements, entries into notebooks, and calculations that he once estimated he had used a million numbers in the entire project.

He wore his shoes, eyes, and patience thin from the constant measurements that he took in the 1820s. The job would occupy ten years of Gauss's life, demanding weeks and months of travel every year. Even though trains would eventually change the nature of travel in Europe during his lifetime, they still had not done so in the 1820s. He spent countless hours in slow, rickety coaches bumping along dusty, rutted roads traveling from town to town and triangle to triangle. The hotels were poor, so Gauss stayed night after night in the uncomfortable lodgings of rural nineteenth-century Germany. He also had to rely on much less intelligent assistants to carry out the work, which meant that he had to be involved in it himself.

The structures Gauss needed in some places to make his measurements were nonexistent, and everywhere there were trees in the way. Trees were a huge problem because there was not always a clear line of sight between two given points. However high a vantage a particular landmark might provide, there were often massive trees obstructing the view. Gauss had no qualms about chopping them down to clear the way for his mapping efforts, but this was not always easy because of the owners or tenants of the land where the trees grew. Gauss once got into a dispute with the owner of a fruit orchard who refused to cut down his trees. How sad yet amusing a snapshot of one of humanity's greatest minds—Gauss snarled in a battle over a few trees, his greatest adversary a farmer.

Gauss spent only a few years making measurements, but by some estimates he wasted decades making sense of all the information. He and his assistants collected more than a million pieces of data, and Gauss used his method of least squares to improve the measurements. In the days before computers, the reduction of observed data took him countless hours, years even. Some consider these lost years, a compelling argument considering the work he might have done had he not been so busy with his surveying. The problems of making sense of the measurements were really a minor theoretical challenge to someone like Gauss.

Some would say that it was a shame a genius as great as Gauss would waste so much time making menial measurements, performing rudimentary calculations, and compiling the pedantic data of a land survey that would be little more than a historical footnote.

In the 1820s, when he might have been hard at work on non-Euclidean geometry, he largely ignored the subject—aside from the already mentioned calamitous exchange with Taurinus.

Still, Gauss was probably not bothered by the rudimentary nature of the calculations. He once said, "Whether I apply mathematics to a couple of clods of dirt, which we call planets, or to purely arithmetical problems, it's just the same; the latter have only a higher charm for me."

Besides that, an argument could be made that the surveying was one of the most important things Gauss ever did. He threw himself into it, and if anything, thanks to his invention of the heliotrope, he could make geometric measurements on a scale that had never before been imagined. In this way, he did make one attempt to test one of the ideas of non-Euclidean geometry—that the angles of a triangle do not always add up to 180 degrees.

On a clear day there are no longer any limits to the sides of a triangle, Gauss wrote in 1825, "except such as are set by the Earth's curvature." He measured a triangle formed among the mountains Inselberg, Brocken, and Hohenhagen, where a tower in his honor was eventually erected. In doing so, he tested one of the ideas of non-Euclidean geometry. According to regular "flat" Euclidean geometry, the three angles of a triangle sum to 180 degrees. This is in fact one of the first things that Euclid proves in the *Elements*. But one of the strange results of non-Euclidean geometry is that the three angles in a triangle sum to less than or greater than 180 degrees. Another strange result is that this deviance is greater the larger the triangle.

While the Inselberg-Brocken-Hohenhagen triangle was certainly huge in terms of the survey, it was only a slight triangle in absolute terms—as measured by comparing its size to the size of Earth as a whole. So even though the triangle's sides were miles long, it was still more or less flat. To an accuracy of about 0.0002 percent the triangle was Euclidean.

Still, the survey work brought Gauss to some of his greatest discoveries in geometry—something he hinted at in a letter he wrote in 1824 to his friend Bessel. "All the measurements in the world do not balance one theorem by which the science of eternal truths is actually advanced," he said. Within a few years, he would write two serious theoretical works based on his experience with the measurements and basically lay the theoretical groundwork for the field of geodesy, a branch of science that deals with mapping Earth. His techniques remained in use for decades.

Gauss created an entire new field with his geometric theories related to curved surfaces, which became the basis for later, even more important work in advanced geometry. This later work would then feed into the mathematical basis of Einstein's theory of general relativity. Einstein once said that if Gauss had not invented the geometry of curved surfaces, probably nobody would have.

Instead of representing a surface in three-dimensional space, Gauss began to represent two arbitrary surfaces superimposed on each other. He discovered that curved surfaces have intrinsic properties, some of which were unchanged by distorting them. In doing so, he came up with the theory of conformal mapping, an important development for the future of geometry. It is a mathematical technique used to convert (or map) one mathematical problem and solution onto another. His work would enable the fundamental measurements of the curvature of Earth and find applications in other fields, such as astronomy. All told, it seemed one of the most exciting prospects Gauss had ever faced in his life.

This was a difficult time at home for Gauss because his wife, Minna, became increasingly ill. In 1822 and 1824, officials in the city of Berlin began seriously wooing Gauss to move there and take a well-paid position as permanent secretary of the academy of sciences there. Berlin at the time was the capital of Prussia, the most prestigious of the German states. Minna probably would have preferred to go there, and it would have been completely natural for Gauss to go to Berlin. After all, for the previous century, some of the greatest minds in Germany and the surrounding countries had gravitated there. Leonhard Euler, the greatest mathematician of the mid-eighteenth century, had gone there at the invitation of Frederick the Great. So had Joseph-Louis Lagrange. Both had produced their greatest bodies of work there.

Minna became enthusiastic to move to Berlin for other reasons. It was a cultural capital with much more to offer socially. She naturally encouraged her husband to make the move. Many of his friends were looking forward to it as well. It seemed to them to be completely natural, and there was no better time than in the 1820s. Gauss, at midlife, was at the peak of his career.

But the negotiations designed to draw Gauss to Berlin were slow and protracted. In the end, the Berlin officials made Gauss a nice offer, and he might have taken it had it not been for the Göttingen officials. They offered him a generous cash bonus and raised his salary to reflect the fine work he had done on the survey and to counter the offer of a better-paying position in Berlin. This counteroffer did not exceed but merely matched the offer made by Berlin, and in the end Gauss may have stayed put for no better reason than to avoid any disruptions in his life. He had lived in Göttingen for the previous two decades, had prospered there, and seemed happy to remain.

Pity poor Minna, though. For her, *staying* was the greater disruption. She might have enjoyed moving to the cultural jewel of Germany. She had suffered in her life. Not that her marriage to Gauss was the wrong move. It was logical and proper, but while it advanced him socially, it didn't do as much for Minna. She would have done better had they moved to Berlin. Perhaps she could have finally spread her wings and soared within greater social circles. Instead the family stayed in Göttingen, in the cold new observatory just outside of town, and Minna remained there until her death.

Minna suffered from bouts of depression and failing health, due, most likely, to tuberculosis. She became more ill after her third pregnancy in 1816, right around the time the old matron, Gauss's mother, moved into the household. After that, she became less and less able to follow the ordinary routines of her life. Even as Gauss was increasingly drawn away from home, toward the German countryside and his triangles, Minna was increasingly bedridden. She would lie in bed for hours as the children played and Gauss worked, increasingly away from the observatory on his long mapmaking surveys, and the sound of her mother-in-law feeling her way blindly through the observatory was a constant presence. The old woman would not die, but Minna certainly would. That was becoming more and more apparent.

Meanwhile, several hundred miles beyond the hills Gauss was surveying, his old friend Farkas Bolyai continued to try to prove the fifth postulate. Farkas's son was about to discover the same passion.

11

A Strange New World

Many things have an epoch simultaneously when they are found in several places, just as the spring violets come to light in several places.

—Farkas Bolyai

János Bolyai preferred to take his final school exams in the summer of 1817, and he passed them easily. He was then ready for the next stage of his life. It was a good time for him to leave home. Things in his household had gone bad, and now they could not have been worse. His father, Farkas, fought constantly with his mother.

The family's financial situation was poor to begin with, but it deteriorated when the local currency was devalued to less than half its value in 1817. Farkas's salary was stretched thin, and thin as it was, he didn't even always get paid on time. János had to find a sponsor to help him pay his tuition. His mother, adding to his troubles, may have suffered deep mental problems. In just a few years she would die, and his father would soon remarry. János's new stepmother was twenty-two years younger than Farkas and only a few years older than János.

It was from this chaotic home front that János Bolyai happily fled to Vienna. There he enrolled in the officer's program at the Imperial and Royal Academy of Military Engineers. Having grown up in a time of military tension, coming of age just after the culmination of the Napoleonic Wars, it was only natural for him to become a military engineer.

János cut an impressive figure as a military man. He was undoubtedly one of the most brilliant students in the school. He had already learned to read multiple languages and do advanced mathematics at a very young age. At the engineering school, he would continue to excel, and he had the makings to be one of the best soldiers in the Austrian army.

He was a dashing soldier and made quite a splash in society—young, brilliant, and handsome in his crisp uniform. Besides being intelligent, he was the most talented swordsman in the army and one of the best dancers as well. He could recite a hundred complicated equations by heart, converse in nine languages, including Chinese and Tibetan, and play haunting tones on the violin. As a student, he was once challenged to a sporting sword duel by thirteen different calvary officers. He agreed to fight them all on the condition that after every match he be allowed to play the violin. He beat all thirteen.

Despite his brilliance in all these ways, János had one fundamental flaw that made him a poor soldier, however. In many ways, his creativity and independence of spirit, which made him a great mathematician, also rendered him fundamentally unfit for military service. He was aware of but immune to the concept of a strict hierarchal system that had to be obeyed without question. It is ironic that someone whose work would overthrow thousands of years of fixed tradition would arise out of such a strict dogmatic profession.

At school, János Bolyai befriended a man named Carl Szász, a fellow Hungarian and another brilliant student. Together they developed an interest in the fifth postulate and began to think about how to prove it. They sought to solve the mystery of more than two thousand years in much the same ways that had been tried by so many generations of mathematicians before them. By then, this had become something of a rite of passage for young mathematicians, and János and Szász started out following the same narrative. They would come to discover the subject in school. They would try to prove it. They would fail. And they would go on with their lives. In a way, the camaraderie between János and Szász somewhat mirrored that of Gauss and János's father, Farkas, a generation before—though Szász was no Gauss, of course. In these early years of his schooling, there were probably few things that excited János more than his conversations with Szász.

In 1821, Szász left to take a teaching position in another city. With his friend gone, János was left to consider the problem for himself. In 1822, he was about to graduate and be unleashed on the world. He would serve as an Austrian officer for the next eleven years, and just as he was about to embark on his tour of duty, he was also about to bring twenty centuries of fruitless efforts to an end by inventing non-Euclidean geometry.

At the outset of his efforts, János wrote to his father and told him what he was working on, letting him know that he was interested in tackling the same problem on which old Farkas had spent miserable decades.

Farkas was horrified to discover that his son, just coming of age, was about to follow in his footsteps and toil away on the monstrous and enormous task of proving the fifth postulate. "You must not attempt this approach to parallels," Farkas wrote to János. "I know this way to its very end. I have traversed this bottomless night, which extinguished all light and joy in my life. . . . For God's sake! I entreat you, leave the science of parallels alone."

In the same letter, Farkas claimed that he risked his life and happiness trying to prove the fifth postulate. "I turned back when I saw that no man can reach the bottom of this night. I turned back unconsoled, pitying myself and all mankind." Learn from my example, he told his son. "I wanted to know about parallels, I remain ignorant, this has taken all the flower of my life and all my time from me."

Young János disregarded his father's advice and plunged into the subject. He gravitated toward the same path that Giovanni Saccheri took a century before, trying to prove the fifth postulate by assuming it was not true and then examining the implications of a system of geometry. It was the same path that Johann Lambert followed but not far enough. It was what F. C. Schweikart understood but couldn't communicate. It was what Gauss knew but never published. It was what F. A. Taurinus wrote but never followed through.

János took a great jump forward, however, because he made the important conclusion that the failure to prove the fifth postulate arose from the fact that it could not be proven. He realized that the alternatives were not contradictory but were the basis of a generalization of geometry. János discovered the first complete non-Euclidean geometry. He eventually called this the "science of absolute space."

János was still not sure at that point what he had discovered, but he was sure that it was something spectacular. On November 3, 1823, he wrote his father a hasty note telling him about his discovery. "I intend to write, as soon as I have put it into order, and when possible to publish, a work on parallels. I have discovered such magnificent things that I am myself astonished at them. It would be damage eternal if they were lost. . . . Now I cannot say more, [except] that from nothing I have created a wholly new world. All that I have sent you compares to this only as a house of cards to a castle."

Some call it the greatest revolution in mathematics since the time of the Greeks—the discovery that changed the definition of a straight line. In ordinary, flat, Euclidean geometry, a straight line is exactly that—its "straightness" defined by the singular unbending direction that it follows for its length. In non-Euclidean geometry, a straight line is defined simply by the fact that it joins two points within a given space. Straight lines may actually be curved. János Bolyai, who called himself Euclid's phoenix, defined absolute geometry as that form of geometry in which the theorems were true regardless of whether they were Euclidean or non-Euclidean.

In early 1825, János was at his peak. The young man was a dashing officer and a brilliant musician. Having come to appreciate and understand non-Euclidean geometry, he wanted to get the opinions of others, so early in 1825 he presented his father with a sketch of his work on non-Euclidean geometry. While Farkas did not completely agree with the work, he nevertheless supported his son's efforts and encouraged him to write it out in Latin and prepare it for publication. He warned János to publish quickly or risk the danger of someone else scooping him by publishing first and establishing priority. "When the time is ripe for certain things, these things appear in different places in the manner of violets coming to light in early spring," he wrote.

That same year, János wrote to a former teacher named Johann Walter von Eckwehr. In the letter he described the major components of absolute geometry as he saw it then. Von Eckwehr was an important man—later the imperial royal general. But he had little time for things he didn't understand. János got nothing back from him. This was the first in what would be a lifetime of disappointments.

• • •

Nikolai Lobachevsky, meanwhile, had made great strides of his own. In 1816, he had just been made a professor at the University of Kazan during one of the toughest periods of the young institution's history. His lectures are said to have been detailed and wonderfully clear, but the outlook for professors at the university could not have been more dismal.

The times were changing. An anti-Western sentiment that had been growing in some of the highest levels of the Russian government came to a head the end of the decade. Prince A. N. Golitsin, the head of the Ministry of Religious Affairs and National Education, was about to take an approach to education that can be described as ferocious.

One of Prince Golitsin's key lieutenants was a tough bureaucrat named M. L. Magnitsky, who was the chief architect of the shake-up that ensued at the university. A few years into Lobachevsky's tenure as professor, Magnitsky prepared a report on the university that was brutal and scathing. Filled with distortions, falsehoods, and innuendo, it ended with the recommendation that the university be destroyed. Not content to wait for anyone to act on his final recommendation, Magnitsky began to see to this personally.

Magnitsky became guardian of the Kazan educational district, and one of his mandates was to establish order at the university. The Ministry of Religious Affairs and National Education had begun a campaign to eliminate Western influences in Russia several years earlier. This was something like a war on philosophy and scientific reason. The University of Kazan was caught up in this attitude as was nearly everyplace else, and Magnitsky was the grand inquisitor.

Magnitsky arranged the dismissals of several professors because they identified with what he saw as subversive philosophical teachings and beliefs. His chief complaint was that these professors taught and studied the works of Immanuel Kant, whose humanism he found threatening. The University of Kazan had been one of the central strongholds of Russian Kantianism, because when it opened its doors to foreign professors shortly after opening, it welcomed several from Germany where Kant was king. Lobachevsky's colleague Bonner, for instance, a professor of physics at Kazan, was a great admirer of Kant. Magnitsky moved almost immediately to dismiss him and several other professors, sending all of the foreign professors home. The university now faced a real crisis, because these professors were among its greatest. They brought a flavor of the rest of Europe to the remote institution. Now they were gone.

In less than a decade, Magnitsky managed to nearly wreck the university's morale and research. "Never before had the university presented such a wretched and shameful spectacle, and great, indeed, was the responsibility of [Magnitsky and others] who brought it to such a state," wrote a historian of Kazan University.

Lobachevsky was the head of the physics department at the time, and he managed to stay clear of the purges. As a homegrown product of Russian education, he was not a target. Still, the purges placed a heavy burden on him. He had to teach all of the courses in mathematics practically by himself. This burden had a silver lining, though, because for one thing it took all of his time and shielded him from becoming too heavily involved in the convoluted politics of the Magnitsky reign of terror. It also made him a leading expert in geometry—so much so that he began drafting a major work in geometry called *Geometriya*.

By 1823, Lobachevsky had finished his book. In it, he began to examine the question of the fifth postulate. "No vigorous proof of this truth had ever been discovered," he wrote. "Those proofs which had been suggested were merely explanations and were not mathematical proofs in the true sense." He included three proofs of the fifth postulate in the book, but he knew that none of these actually did the trick. He admitted as much himself, writing that nobody had adequately proved the postulate. Such expressions of self-dissatisfaction may have doomed his book.

Lobachevsky gave his book to Magnitsky in 1823, and Magnitsky forwarded it to a professor named Fuks. He didn't tell Fuks who had written it and asked him to evaluate its worth as an elementary textbook on geometry. Fuks was a professor of medicine and the rector of the university. He had been so put upon by Magnitsky that he completely lost his will and became what one historian refers to as a "blind instrument" of Magnitsky's petty cult of personality. He sensed that Magnitsky wished him to judge the book poorly, and so he acted accordingly. He dismissed its worth, and attacked it for containing ideas influenced by the French Revolution. It was never published.

In a way, Lobachevsky was probably okay with that. By the time the book was being evaluated, he had already decided against his own work. The setback of being judged harshly by a reviewer was presumably softened by the fact that Lobachevsky had already come to the same conclusion himself—that it was not worthy of publication.

No historical record exists as to why he did what he did next, but for some reason he set out alone in the figurative wilderness. He was headed for an uncertain future toward non-Euclidean geometry. Lobachevsky threw himself into an intense concentrated effort to understand the fifth postulate. Because of the attack on Kant, the mood was ripe for Lobachevsky to attack the notion of Kantian space. He came to consider the fifth postulate in depth and slowly began to convince himself that it might be possible to have a completely sufficient system of geometry in which the fifth postulate was not true. In two years this single man managed to make more progress than humanity had in the previous two thousand.

Lobachevsky spent many months developing this geometry, and he began to consider the possibility that two forms of geometry could exist in the world, one governed by physical forces and another by logical consistency. He gave his work to three of his colleagues to review, but he never received a response from any of them. Finally, by early 1826, he was ready to present the material to the scientific society at Kazan.

This was a time of abrupt changes. The campaign of the Ministry of Religious Affairs and National Education began to falter, and many new ideas in literature, philosophy, and science would be introduced and embraced. It was an exciting, open time.

The Magnitsky reign of terror ended suddenly when Czar Alexander died. Alexander was succeeded by Nicholas I, and Magnitsky had written some nasty memoranda about Nicholas over the years. He thought so little of Nicholas's chances of becoming the next czar that he traded diplomatic caution for the expediency of his little polemics, which he sent to Alexander. After Alexander died, though, Nicholas got his hands on some of these documents. In them Magnitsky accused him of being too liberal. Nicholas was furious.

Magnitsky was unceremoniously dumped from his position as guardian of the Kazan educational district and exiled. He was replaced by Count M. N. Musin-Pushkin, a former student at Kazan. Musin-Pushkin inherited an institution in trouble. Many of the best members of the faculty had been dismissed, and those who remained had long been terrorized during Magnitsky's reign. Morale was nonexistent. Musin-Pushkin needed to do something to repair the damage. So one of the first things Musin-Pushkin did was to appoint Lobachevsky the new rector of the university in 1827, replacing Professor Fuks, who had derailed the publication of his geometry book a few years earlier.

Around the time János Bolyai and Lobachevsky were inventing non-Euclidean geometry, Gauss, fresh off his run-in with Taurinus, was beginning to turn again toward geometry. This time he was interested in the theory of curved surfaces—something influenced, no doubt, by his many toils measuring triangles on the curved surface of Earth.

The mapping work reinvigorated Gauss's interest in geometry. The greatest unsolved problem at the time was still that of the fifth postulate, but this mystery was by no means holding back the field from new developments. Geometry had been moving in different directions for decades, and Gauss was becoming interested in these more recent discoveries.

Decades before Gauss was born, Leonhard Euler had developed differential geometry, which extended geometry to include the use of differential calculus. This work was carried further by Adrien-Marie Legendre. Projective geometry, which deals with geometric bodies under projection, was then developed in the nineteenth century by an officer in Napoleon's army named Jean-Victor Poncelet. He was captured by the Russians during the disastrous Russian campaign and conceived of projective geometry while serving time in a remote Russian prison. One of the big problems that this form of geometry addressed was the development of methods to translate three-dimensional shapes onto planes in the most accurate way. This conformal mapping can be used to project the surface of a sphere onto a flat map accurately.

Gauss became interested in all of this while he was doing his mapping, and he proposed a public challenge, called the

Copenhagen prize, which sought geometrical solutions for producing projections onto maps. When nobody successfully came forward to solve the problem and claim the prize after a few years, Gauss solved the problem himself and advanced the field significantly.

In 1827, Gauss began to work intensively on the foundations of geometry. He was coming at this from an intimate understanding of the relationship between geometry and our curved Earth. We are small beings on a huge planet. To us the world seems flat. But our planet is curved, and this influences the geometry of objects mapped on it. Gauss began thinking of the intrinsic geometry of a surface. He picked up the work of Euler and derived a new way of doing this conformal mapping through a particular type of operation known as transformations. With these transformations Gauss could map parts of a three-dimensional space to a flat space. His work set the standard for all future work in the field.

What Gauss realized is that any surface has an intrinsic curvature that remains unchanged even if you map that surface onto another surface. The curvature comes not from the thing drawn but rather the surface upon which it is drawn.

The easiest way to picture this is with an inflatable playground ball. Empty the ball of air and it becomes flat—but not perfectly flat. Its ends are still curved and it may have a bulb or a curved indentation in the middle. In other words, it retains its curvature even though it is no longer spherical. Now imagine that in your zeal you want to really flatten the ball, and you pop it, tear it apart, and spread its skin flat on the asphalt. When you flatten the pieces of the ball, you will find, again, that they do not lose their curved shape unless you stretch the material.

Gauss was able to study the nature of curved surfaces, asking questions about the properties of the curvature and how these properties were changed (or invariant) when deformed. His idea was that the curvature of the surface was intrinsic to it and that it could be measured by taking into account only the surface without thinking of how the surface sits in three-dimensional space. Gauss worked out the measure of curvature—now called the Gaussian curvature—as that invariant quality of curvature that remains unchanged as you flatten the ball. Thinking about this helped him develop what he called his exceptional theorem (*theorema egregium*), which basically says that if you move a curved surface without stretching it, the curvature remains the same.

This inspired Gauss to write out his thoughts on the general theory of curved surfaces, and within a few years he had finished a major work on the subject. One of the most important books of his lifetime, the *Disquisitiones Generales circa Superficies Curves* worked out the general formula for measuring curvature and established the basis for the modern theory of curved surfaces. It laid the groundwork for advanced geometry.

Gauss was already famous and well respected before all of this later work in geometry. But the widespread utility of this work made his genius more accessible to the general public.

Unfortunately Gauss's mapping efforts did not endure as long. The tragic epilogue to the survey was that aside from the theoretical work Gauss did in revolutionizing the field of geodesy, very little remained by way of legacy. Landmarks disappeared, and after he died, most of the physical points of reference were gone.

12

A Message for You, Ambassador

If two circles touch one another, they will not have the same center.
—Proposition 3 from Euclid's *Elements*

When a massive cholera epidemic spread through Russia in 1827, it was the continuation of an earlier pandemic that originated in 1826 in northeast India's Ganges River delta and migrated east and west from there. The illness traveled up the Ganges into Punjab and spread along whatever routes humans take. By 1827, it had reached the cities of Lahore and Kabul. It crossed over into eastern Russia the following year and began making tracks westward.

On August 26, 1829, what was seen as a mysterious illness was discovered in the river town of Orenburg, Russia. The local medical board took two weeks to diagnose the illness as cholera, and by then the disease had spread. It was soon too late as many more cases appeared. The most devastating cholera epidemic Europe had ever witnessed was beginning. By October there were hundreds of reported cases in Orenburg, and while the disease subsided within a few months, Orenburg was a commercial hub. Its roads led to destinations throughout Russia, and along these routes cholera creeped, spreading illness wherever it roamed.

A quarantine was put in place around the city on October 9, but it was too late and not strict enough. The disease leaked out. The ensuing cholera pandemic had a 90 percent mortality rate in some cities. More than half a million people were infected, and hundreds of thousands died. In Tblisi, Georgia, two-thirds of the city's inhabitants fled in fear.

By 1831 Russia's main cities were infected, and still the disease kept coming. It left Russia and moved into Germany, Finland, Sweden, and Vienna. It spread to North Africa, Spain, Portugal, France, and Britain, and Irish immigrants carried the bug to the United States, Cuba, and Mexico, where it killed tens of thousands.

But in Kazan, Lobachevsky was a hometown hero. In the summer of 1830 cholera devastated the city, but at the university, Lobachevsky took charge of the situation and successfully kept deaths low among the students and faculty. The city suffered 60 percent mortality in the cases seen there, but Lobachevsky sought to isolate the university and protect its students and faculty from the ravages of the disease. He established an infirmary and instituted public health measures that improved cleanliness and limited contact. The death toll was only twelve out of six hundred members of the Kazan University community, which was much less than what some municipalities suffered.

Lobachevsky was able to save the day because he actually held the reins of power at the university in the 1830s. He had risen through the ranks to his position of prominence through a combination of hard work and good luck. In 1827, he became rector of the university, and he held this post for almost twenty years, enjoying a stellar career as an administrator.

Lobachevsky sought to improve the university by reorganizing it and uniting the faculty. He worked not only on the university's morale but also on its physical environment. He even studied architecture to this end, and after securing appropriations necessary to fund the construction of new buildings on campus, he personally supervised the work.

When other potential calamitites arose, Lobachevsky did what he could to protect his university. A fire struck the city of Kazan in 1842, and high winds spread it quickly to the university's buildings. Lobachevsky organized fire brigades and exhorted the school's students to fight the flames. He managed to save the major scientific equipment of the university and the library. After the fire, he worked hard to restore what was lost.

Lobachevsky became a luminary of the university and the town. He organized classes for adults that were not part of any curriculum or degree program. He created a trade school. He helped found an orphanage. Lobachevsky became popular with the townsfolk for delivering lectures on science and for teaching elementary lessons to them, and he was favored within his more exclusive circles because he sometimes threw great parties.

In a sense, the cholera action solidified Lobachevsky's respectability as a dean, but he had already long since established his position as a brilliant academic—even if few appreciated just how brilliant. His greatest achievement was to elevate logical supremacy over intuitive understanding. He also independently discovered non-Euclidean geometry, and he wrote about it before anyone else.

What Lobachevsky did was one of the bravest intellectual acts in the history of mathematics. He stood at the edge of the cliff, just as many had done before, but instead of turning back or trying to find a way around, he plunged into the depths and discovered the secret that had eluded so many for thousands of years. He survived the fall and came back to tell all. His invention—non-Euclidean geometry—challenged the conventional notions of space and geometry. It was as if he had climbed the mountain to find a way down the other side and returned instead to claim that he could wave his arms and fly.

In pre-cholera Kazan, the tall, gangly, soon-to-be-appointed rector Lobachevsky stood in front of an assembly of his colleagues, the university physics and mathematics faculty. There he read a paper that was the fruit of several years' effort on geometry and the fifth postulate. The title of this talk was "A Concise Outline of the Foundations of Geometry with Vigorous Proofs of the Theorem of Parallels." The event, which took place on February 23, 1826, is sometimes called the birth of non-Euclidean geometry.

Lobachevsky became interested in the fifth postulate in 1815, and for the next seven years he worked on the problem on and off, even as the university was going through the turmoil of losing all of its foreign professors. Like all mathematicians before him, Lobachevsky started out trying to prove the fifth postulate. Failing, as many before him had, he overcame his frustration by investigating what would happen if it were not true. He began to see space not as the necessary realm involved by philosophers but as something our minds create—a convenience to which we subscribe because it conforms to the world we see.

For Lobachevsky, embracing the idea that the fifth postulate was not true did not introduce a contradiction but rather led him to a completely logical new type of geometry. He called this his "imaginary" geometry, and after his death, some would call it Lobachevskian geometry. Lobachevsky replaced the fifth postulate by proposing that more than one straight line through a single point could be drawn parallel to a given line in the plane. In fact, in his new geometry, there were an infinite number of parallel lines. He accepted that instead of parallel lines never approaching one another, in non-Euclidean geometry, lines continually converge.

The shortest path joining two points in non-Euclidean geometry is not a straight line in the sense that we know it. The easiest way of imagining this is to look at a triangle in the geometry that Lobachevsky invented. Triangles have curved sides. They curve inward toward the center of the triangle, and this effect is greater the larger the triangle is. The three angles of a triangle are equal to 180 degrees according to Euclidean geometry. But according to Lobachevsky's imaginary geometry, the sum of the three angles must be less than 180 degrees. One of the strangest results is the "lost rectangle." Rectangles don't exist because it is impossible to draw a shape with four right angles.

This strange warping space of non-Euclidean geometry was striking, impossible, and completely new. In regular geometry, three points in the space either define a straight line or they define a circle. In non-Euclidean geometry, three points can exist that are neither on a line nor forming a circle.

Lobachevsky was well aware of the implications of his discovery. He wrote, "The futile efforts from Euclid's time on throughout two thousand years have compelled me to suspect that the concepts themselves do not contain the truth which we have wished to prove, but that it can only be verified like all other physical laws by experiment, such as astronomical observations." The reason he suspected that one could make astronomical measurements to verify the geometry of space was that the curving nature of the sides of a triangle became more pronounced as the size of the triangle became larger. Gauss had realized this as well, which was why he tried to measure the Inselberg, Brocken, and Hohenhagen triangle. This was a very small triangle in terms of Earth, encompassing only one small corner of Germany. In terms of the solar system, it was a mere speck.

Lobachevsky's idea was to measure a much larger triangle indeed. He knew, however, that the deviation would be greatest where the triangles were the largest. "Nature itself points out distances to us compared with which even the distance from the Earth to the fixed stars disappears to insignificance." He attempted some astronomical measurements to test whether his geometry would be the best type to represent the space of the universe. He looked at the angles of a triangle made by Earth, the sun, and Sirius, a nearby star. He found, however, that the angle sums were more or less consistent with Euclidean geometry, and he concluded that Euclidean geometry is far-reaching.

Nevertheless, when he was finished developing his imaginary geometry, Lobachevsky knew that he had created a new epoch—much as János Bolyai realized this a few years before him, and just as Gauss probably guessed a few years before both of them. Suddenly space was not the way it had been for thousands of years. But nobody seemed to care.

Not a single colleague of Lobachevsky ever expressed a favorable opinion of his work. Nor, for that matter, did anyone else in Russia. He submitted his accompanying paper for review and deposit in the university's archives, but it was not well received. It was put aside for eight years until it was finally deposited in the archives, and then it was duly lost.

A few years later, Lobachevsky published his first article on the subject in installments in an obscure journal called the *Kazan Messenger* in 1829 under the misleading title "Elements of Geometry." He followed this up with several more papers in the coming years. In doing so, he beat young János Bolyai by a few years.

Kazan University was remote. It was no European intellectual capitol, where there might actually be people who could understand Lobachevsky's work. So having failed to get anyone to pay attention to his work there, Lobachevsky sought a larger audience elsewhere. He submitted his work to the St. Petersburg Academy of Sciences. There it came to the attention of Mikhail Vasilievich Ostrogradskii, one of the greatest Russian mathematicians of all time.

The scion of a wealthy and noble family, Ostrogradskii was a throwback to the time when many great Russian things came in a western European wrapper. He availed himself of the best education available to him at home, and then he migrated to France and was educated in Paris, where he made a splash in scientific circles before returning to Russia. He was so connected to Paris that when he returned, he was initially placed under surveillance by the secret police, who were probably acting out of the same sort of anti-Western sentiment as the Ministry of Religious Affairs and National Education in its campaign to cleanse Mother Russia of the European Stain.

But Ostrogradskii grew to prominence despite this surveillance. He rose quickly through the ranks in St. Petersburg and began to produce an impressive body of work. By the early 1830s, he became something like an ambassador of Russian science. In Russia, he was without apparent peer. He was at the very center of mathematics there—perhaps even for the next few decades he *was* Russian mathematics. He brought prestige to the discipline in general and attracted students from far and wide.

One of his duties was to make important decisions on who would be appointed to which academic positions within mathematics and other fields. This had the effect of elevating his authority beyond mere influence and into the realm of cult of personality. Ostrogradskii also selected which studies would be published in the journals of the St. Petersburg Academy. It was in this role that he would exert vast power over Lobachevsky.

Ostrogradskii was fast on his rise to greatness when Lobachevsky's work came across his desk. He was perhaps the one person who might have been able to influence popular opinion in Lobachevsky's favor had he chosen to highlight the paper and help to popularize it. He chose otherwise, however.

Ostrogradskii rejected the paper as unworthy for publication in the St. Petersburg Academy of Sciences. From his point of view, the work was strange and unbelievable.

Intellectually, Ostrogradskii was concerned with mathematics of old—bodies in motion, elastic collisions, fluid mechanics, heat, and acoustics. These were the sorts of areas of applied mathematics that had strengthened the case for Euclidean geometry, and they were the types of applied real-world problems that Ostrogradskii studied. He was a mathematician of the most applied sort.

One of the main problems with non-Euclidean geometry as Lobachevsky presented it was that it lacked any tangible application. Because Euclidean geometry seemed to hold for the sort of distances that one might be concerned with in the real world, the value of the new non-Euclidean geometry was suspect to someone like Ostrogradskii. Lobachevsky could not present any use for it, practical or theoretical. Besides that, he called it "imaginary" geometry. But geometry was the mathematical science of the real. How could Ostrogradskii ever consider taking it seriously?

In a sense, ordinary Euclidean geometry was the basis for everything that Ostrogradskii had ever done. It was a system that worked, and for him there was probably no reason to even question it.

If Ostrogradskii's opinion of imaginary geometry was not kind, his regard for its author was no more polite. Lobachevsky was on the periphery of Russian science. He was the poor son of humble parents, not a privileged child of a famous family like Ostrogradskii. Nor was he educated in France. Nor was he living in the metropolis of St. Petersburg—or even Moscow for that matter. Lobachevsky was a homespun mathematician from the academic outpost of Kazan. So Ostrogradskii had no motivation to do anything but what he did, judging the work to be unworthy for publication and rejecting it outright.

To add insult to injury, Ostrogradskii also was the presumed author of an anonymous review of the paper in a St. Petersburg journal titled *Son of Fatherland*. In the review, he skewered Lobachevsky, dismissing his work as a troubling combination of falsehoods, trivialities, and indecipherable discourse. He stated that what was new was false, and what was not new was trivial. Overall the work was unintelligible, the review said.

Hurtful as it sounds, this last bit of criticism may have been the most constructive. Ostrogradskii would not be the last person to note the impenetrable and opaque nature of Lobachevsky's writing style. Perhaps this was what Lobachevsky needed to hear. He knew his discoveries were significant—not only because they solved the mystery of the fifth postulate, which had persisted for more than two thousand years, but also because his imaginary, or non-Euclidean, geometry was a fundamentally important development in mathematics. It separated mathematics from truth and allowed future generations of mathematicians the ability to spread their wings. Lobachevsky knew this. He had just failed to communicate it.

So Lobachevsky marched on with trying to gain recognition for his work. Undeterred by the setback of having his work rejected by the St. Petersberg Academy and skewered by Ostrogradskii, he began to rewrite his work with renewed vigor. That was in 1831, the year when the cholera epidemic struck.

That same spring of 1831 János Bolyai went to visit his father. He was in the middle of transferring to a new military post in the city of Lemberg (now Lviv, Ukraine), and on his way there, he visited his father in Marosvásárhely. Farkas was about to publish his magnum opus, a two-volume tome called the *Tentamen*, which was an outstanding summary of the mathematics of the age. Farkas convinced his son to include his work on non-Euclidean geometry, now a few years old, in the appendix of the *Tentamen*. Had his father not done this, János later wrote, the science of absolute space "would never have seen the light of day."

János's work has been called the most extraordinary two dozen pages in the history of thinking. It was merely an appendix, but it became more famous and far more important than the rest of the massive two-volume work in which it appeared. Still, it was ultimately a disappointment to János. He didn't make any money off the treatise. In fact, he had to pay out of pocket for the printing.

Money aside, the greater debt the work incurred was one of sorrow. János had never published anything in his life. Getting his work in print was a dream come true and more. It meant that he could now be recognized as the mathematical genius that he was. All indications are that János fully expected this recognition to come eventually. Publishing put his expectations on the boil. But success never came. János was not recognized for the work in his lifetime.

The ultimate insult was that János was not even the first person to publish on this subject. He had been beaten by a few years by Lobachevsky. Though he would not discover this fact for many years, he would be so disheartened from being beaten by an obscure Russian mathematician writing in an even more obscure journal that he refused to believe it.

These disappointments were still years away in 1831, though. János was still excited then, and he looked forward to greeting the greater world of intellectuals that he presumed his work would introduce him to. Farkas, for his part, was his son's most enthusiastic supporter—not only because he published the appendix but also because of what he did with it. Farkas sent the *Tentamen* to Carl

Gauss in April 1831 and followed this up with a letter in June. He was maximizing the possibility of getting his son's work recognized by sending it to the greatest mathematician of the age.

The year 1831 was particularly chaotic for Gauss. The cholera epidemic had struck Göttingen as well, sickening some of his workers. Besides that, his wife was now almost completely bedridden.

Against this backdrop of disease, some of the local residents of Gauss's city were agitating against the Hanoverian dynasty, which ruled the city from the throne of Great Britain. This arrangement had existed since England's king George I, formerly the duke of Hanover, rose to the throne of England more than a century before. For more than a hundred years, the ruler of England had also ruled Hanover and the surrounding areas. The situation was soon to change, however, thanks to a popular youth movement.

Enrollment at the university in Göttingen had increased dramatically, and the students were emboldened by the spirit of rebellion in Europe. A few weeks after the Christmas break, the students stormed city hall and dissolved the city council. They installed their own council, and they prepared to establish the commonwealth of Göttingen. As revolutions went, however, this one was short-lived and relatively unexciting. The king's army approached the city, and the students quickly moved aside and allowed them to restore order. Concessions were made, though, and a local bureaucrat named Count Münster, who was particularly unpopular with the students, was forced to resign.

Gauss could not have been more uninterested in the rebellion. "Basically I am little touched directly by the local events," he wrote to a friend shortly after the army ended the short-lived reign of the student force. In the same letter, he mainly details the petty comings and goings of the servants in his own household. He was clearly addled with his own form of chaos.

Perhaps because of the close proximity of his dying wife and with so many people all over Europe perishing from cholera, Gauss was reminded of his own mortality. That year, in 1831, he decided he was ready once and for all to tackle the problem of the fifth postulate. He had already seen what would happen if he rejected it.

Gauss had understood—independently of Lobachevsky and János Bolyai—that the solution to this mystery was to accept that there was no way to solve it. The solution was to examine what geometry would be like without it. He knew he could invent non-Euclidean geometry and that in doing so he would be changing mathematics forever. In 1831, he knew it was finally time for him to write it all out. Surrounded by death, he must have thought that if he died, non-Euclidean geometry would expire with him. That would have been a tragedy for mathematics.

So Gauss began to formulate his ideas, probably planning to publish them in a few years and let the chips fall where they may.

13

"To Praise It Would Be to Praise Myself"

In vain, diverse geometers whom this arrangement has displeased, have attempted to better it. Their vain efforts have made clear how difficult it is to substitute for the chain made by the Greek geometer another as firm and as solid.

—Jean-Étienne Montucla (1725–1799)

Carl Gauss taught his children to reject mathematics. Whether acting out of pride, indifference, concern, mercy, or genuine love, he thought they could never measure up to the standard he set, and he believed they would never have a prayer of surpassing him as mathematicians. They were all smart, and one or two of them even showed signs of brilliance. The issue was more the awareness on Gauss's part of his own unique genius. He saw to it that all his children were well educated, but he forbade them to follow in his footsteps. He wanted to guard against the potential misery they would suffer as the overshadowed children of a mathematical legend.

Whatever strange compassion compelled him, Gauss guaranteed the legacy of his name by keeping his offspring off numbers. This was probably not a difficult task. Becoming a mathematician was a long, hard road. It took talent, hard work, and more. Whatever discouragement his children needed, they probably didn't need much. A gentle nudge was likely enough to ensure that they all had no interest in following in their father's footsteps—all except one.

Gauss's son Eugene was, by all accounts, brilliant—perhaps as brilliant as his father. He showed a keen interest in mathematics and might have even become as talented a mathematician as his father. The boy appeared to have the same combination of mathematical and language prowess. But Gauss would not let Eugene pursue this path. He had formed high hopes for his son's pursuit of a different career early in his life, and he tried to convince him to study law instead. This may have precipitated the difficulties their relationship encountered during Eugene's teenage years.

These difficulties were not solely of Gauss's making. Like many young students who are forced into a field in which they have no interest, Eugene could not focus his energy on his legal studies. Instead, he found other, more extracurricular activities to turn his attention toward. The exact details of all the questionable activities that sustained his early college years have been lost to time, but they can easily be surmised. Eugene was something of a wild man when he got to college. He partied a lot. He even fought a duel. It was this lifestyle that would cause the most disappointment to his father.

Eugene's tenure as a reckless student was short-lived and came to a grinding halt in 1830 when he was forced to flee gambling debts and other troubles in Göttingen. It started one night when he threw an elaborate dinner party for his fellow students and stuck his father with the bill. Gauss was furious at this, and after he confronted his son, Eugene picked up and fled. Gauss followed him and tried to convince him to stay. Father and son had one final meeting in the city of Bremen, but the boy was adamant. He was going to run away to the United States.

Eugene was only nineteen when he emigrated, and he and his father parted on somewhat bitter terms. They never saw each other again. For Minna, this must have been even more heartbreaking. She didn't even get to say good-bye.

Philadelphia, where Eugene eventually landed, was one of the most common points of arrival for German immigrants like him. Tens of thousands of German men, women, and children came, some of them with little more than the clothes on their backs. They moved into urban Philadelphia neighborhoods, and they populated Pennsylvania's rural backwoods. They were farmers, craftsmen, factory workers, hucksters, soldiers, and merchants.

Eugene enlisted in the U.S. Army upon his arrival in Philadelphia. This was a good way to guarantee fast-tracked American citizenship. His plan was then to get an early discharge, which in those days could be purchased, but required more cash than he had. He wrote to his father asking for the money, but Gauss ignored him. So Eugene was stuck. He was posted to Fort Snell, near St. Paul, Minnesota, and served alongside Jefferson Davis. Because of his prestigious father and learned background, Private Eugene was made the fort librarian.

Joseph Gauss, Eugene's older sibling, tried to improve his brother's fortunes by appealing directly to General Winfield Scott, seeking an officer's commission for his brother. This might have happened, too, but Eugene had other plans. He left the army and sought his fortune with the American Fur Company in the headwaters of the Mississippi River. There his linguistic talents came in handy, and he quickly became fluent in Sioux. Then he traveled down the Mississippi to Missouri and settled in St. Charles. He made a good living there as a businessman, was married, and lived a long and good life, raising seven children before settling down as a gentleman farmer.

America was a land of opportunity for Gauss's youngest son, William, as well. In Germany, William too was having a difficult time. He was equally frustrated in his chosen career as a farmer's apprentice, and in the fall of 1837 he left for America with his young wife in tow. They arrived in the port of New Orleans and went by riverboat up to Missouri, where they also settled in St. Charles. Gauss's first grandchild was born there in 1838.

Eugene and William were at the very early crest of a wave of German immigration to the United States. They were really part of a greater trend—an exodus of masses of people from all over Germany (and the rest of Europe) toward a new life in America. Several cities emerged over the nineteenth century as centers of German population and, by extension, culture.

This influx would peak in mid-century, dip in the years surrounding the Civil War, and then rebound stronger than ever toward the end of the century. By 1850 some 2.5 percent of the U.S. population was German-born. Within another decade this population had doubled. By 1890 there were nearly three million German-born immigrants in the United States, including more than one hundred thousand in Missouri, where Eugene eventually settled. This was nearly 10 percent of the population in those days.

All these moves were difficult for the elder Gauss. He was separated by a distance of thousands of miles and an intervening ocean, and he would never meet his American grandchildren. The departure of Eugene touched off bitterness and depression in him the likes of which he never displayed at any other point in his life. This loss was a severe blow to Gauss—probably more so because he felt the need to go through the heavy-handed motions of playing the disapproving father. Shortly after his son left for the United States, Gauss grew almost lethargic and then angry. He called Eugene a good-for-nothing son who had brought only shame to his name.

Eugene eventually burned most of the letters his father had sent him, and he melted down a medal that King George V of England and Hanover had given his father. His objective was to use the gold in the medal to fashion himself a pair wire frames for eyeglasses.

Gauss's name failed to produce great fortune for his oldest son, Joseph, as well, and Gauss was frustrated by Joseph's failure to thrive in his chosen career as an army engineer. The frustration stemmed from the fact that Gauss had tried for several years to use his fame and considerable reach along various political channels to have Joseph promoted through the ranks. These efforts failed, and Joseph wound up quitting the army and becoming a civil engineer for a railroad company. While Joseph prospered in his new position, it moved him out of Göttingen, and for the last two decades of Gauss's life his contact with his oldest son became less and less frequent. Joseph even visited the United States in the late 1930s as a representative of the railroads around Hanover. Torn from the larger part of his family by then, Gauss must have wondered, could Joseph have been planning to emigrate to the United States as well?

Gauss also lost one of his daughters to marriage. In the fall of 1830, his daughter Wilhelmine (who went by Minna, the same name as her step-mother) married a scholar and colleague of Gauss's named George Henry Ewald—an academic who was, like his new father-in-law, from a working-class household. As happy as this occasion was for Gauss, it meant losing his daughter when she left home.

By far the greatest blow that befell Gauss came on September 12, 1831, when his wife, Minna, died. Their marriage might have been born somewhat out of convenience and the need to have a wife after the death of his first wife, Johanna, but he had grown to love her deeply. They had spent twenty years together—some of the richest of his life. They lived together in Göttingen longer than Gauss had lived anywhere else. Minna's death was a major loss. Gauss wrote to a friend a few months after Minna died, "I have lost all my desire and will to live and do not know if they will ever return."

After his wife's death, Gauss became more and more introspective. What remained was his work, but he was less likely to engage in new mathematical work. In a way he didn't need to. He was already a famous man, and there was nothing any further mathematical labors could add to this. He was content to look back—in many ways. He began to grow more fond of reflecting back on his youth. He would relate stories of his childhood to anyone who would listen, and as a famous and respected figure, he found no shortage of eager listeners.

Where once Gauss was content to keep his discoveries in non-Euclidean geometry to himself, in 1831 he began to write down his thoughts. Up to this point, a couple of letters, a few scattered notes, and two short review articles were all that he had ever written on the subject. He wrote to his friend Heinrich Schumacher on May 17, 1831, that he was about to begin writing down his thoughts on non-Euclidean geometry so that they would not die with him.

Shortly thereafter, Farkas Bolyai wrote to Gauss in the summer of 1831 about his son, János. His letter was filled with the praise a proud father would have for his legacy. "My son is already First Lieutenant in the engineering corps and will soon be Captain, a handsome youth, a virtuoso on the violin, a fine fencer and brave," Farkas wrote. He also called his son an impassioned mathematician with very rare gifts of mind. "At his desire, I send you this little work of his. Have the goodness to judge it with your sharp, penetrating eye and to write your high judgment unsparingly in your answer, which I ardently await."

Farkas waited more than six months without hearing anything back from Gauss, so he wrote to him again. He included another version of his son's work, this one showing corrections to the draft.

After a decade of toil, the fruits of János Bolyai's efforts were a definitive early work on the subject of non-Euclidean geometry. The full title of the work was, in Latin, *Scientiam spatii absolute veritam exhibens:*

a veritate aut falsitate Axiomatis XI Euclidei (a priori haud unquam decidenda) indepentem: adjecta ad casum falsitatis, quadratura circuli geometrica (Appendix, the absolute true science of space exhibited: independently of the 11th Euclidean axiom [that can never be decided a priori] being true or false: for the case of being false the geometric squaring of the circle is supplemented). The work became largely known by the more common name *The Science of Absolute Space.*

János had a sense of how important the work was, and he thought that it would have spectacular influence. He believed it would lead to fame and recognition, and perhaps it should have. Years later, his work would be called the most extraordinary two dozen pages in the history of thinking. He would eventually become recognized as Hungary's greatest mathematician for his role in the discovery of non-Euclidean geometry—the man who changed space, he would be called. But this was years away. In the 1830s, nobody knew who he was.

The first to recognize the genius of the work was Gauss. He wrote to his friend Christian Gerling in 1832 that he was ecstatic at reading Bolyai's work. "A few days ago I received from Hungary a short work about non-Euclidean geometry," Gauss wrote. "[In it] I find all my own ideas and results developed with great elegance."

Gauss didn't stop there. He replied to Farkas on March 6, 1832, in a letter he intended to be encouraging but which had the opposite effect. He began with an apology that he could not praise it. "To praise it would be to praise myself," he explained, saying that he had thought about the same things for many years. "Indeed the whole contents of the book, the path taken by your son, the results to which he is led, coincide almost entirely with my meditations, which have occupied my mind partly for the last thirty or thirty-five years."

As a way of offering some bit of helpful advice, Gauss warned that János's work was "in such a concentrated form that it will be hard to follow for someone to whom this matter is foreign." He made one concrete suggestion: that János reconsider his use of symbols in the book. Gauss thought some of the ideas in the book could be better communicated if they were described using words rather than symbols alone.

Gauss knew that this was an important discovery, and he was probably relieved and delighted to read János Bolyai's treatment of it because it meant that he no longer had to publish his own thoughts on non-Euclidean geometry. He told Farkas that he would have written his own version eventually. "My intention was to let nothing be known during my lifetime of my own work, of which moreover until now little has been put on paper."

The reason he had never published anything like this himself, Gauss told Farkas, was that it was too complicated for most people to understand. "Most people do not have the right sense of what is involved, and I have found that only few people who receive with special interest that which I have communicated to them. In order to do that one must have felt quite vividly what is really missing, and most people are in the dark about that," he said.

Gauss was more than happy to see that someone else had done it first—especially the son of his old friend. János Bolyai had essentially saved him the trouble, he said. "And I am very glad it is just the son of my old friend, who takes the precedence of me in such a remarkable manner," he added.

Farkas was pleased that Gauss had showered so much attention on his son. Farkas wrote to János, "Gauss' answer with regard to your work is very satisfactory and redounds to the honor of our country and of our nation."

János, on the other hand, was less enthusiastic when he heard about what Gauss had said. The feedback he gave caused János to become furious with his father. Gauss had said it reminded him of his own work, "of which *until now* little has been put on paper." The implication was stinging. Gauss implied that János's work was not his own. "To praise it would be to praise myself" was a statement of Gauss's priority, and this outraged János. Gauss's insinuation was that the work was not original at all, but merely an echo of what Gauss had himself earlier accomplished.

How could Gauss claim the science of absolute space as his original discovery? Moreover, how could he claim it was something that he had thought about thirty-five years before? Was he trying to steal credit for the invention? Was János's father complicit in betraying his ideas to Gauss so that Gauss could take all the credit for them on his own? János was so annoyed by all of this that his relationship with Farkas became strained. They did not speak to each other for a long time afterward.

Things started going badly for János Bolyai. After a short stay at his post in the city of Lemberg, he was promoted to captain and given a new post at Olmütz (now Olomouc, Czech Republic) in 1832, but his military career hit a serious bump in the road, and on his way to Olmütz, two things happened that would change his life. One was that he got into a nasty argument with a customs official who had stopped his coach and wanted to search the contents of his trunk. János refused and was duly reported to the authorities, who notified his commanding officers. His insubordination was noted and later used against him as justification for discharging him.

The second event was much more severe. János's coach overturned and he suffered a serious head injury. His health at that time was not very good to begin with. He was suffering from a recurrent fever that may have been caused by malaria, and he seems to have suffered from cholera along with much of the rest of the population of Europe in those days.

As if these injuries were not bad enough, János's job had become a burden. As a military engineer he was called upon to draft simple schematics all the time—a type of work that was deeply beneath him. He had, after all, come up with the greatest revolution since the time of the Greeks. Now, with the traumatic head injury, he may have suffered headaches or worse.

This was the new János Bolyai: malarial, with cholera, having sustained a head injury, and terribly distracted from his work by his true interest. Physically and mentally unbalanced, he became more involved with his mathematical work and less interested in the military. He tried to obtain a leave of absence for a few years so that he could concentrate on his mathematics, but his application was rejected and he was forced to take repeated medical exams. He left the army in 1833, and his career never really recovered from this blow. He accomplished little in the remaining three decades of his life.

The next few years of János's life were as chaotic as they were depressing. He moved in with his father in 1833 but moved out a year later. He then moved to a family estate that Farkas had inherited from his mother. János met and fell in love with a woman named Rozália Kibédi von Orbán, and they lived together and had a few children. They could not get married, however, because János had trouble managing his money, and it took money to get married in those days. An obscure regulation governing military officers would have required him to pay a hefty deposit. Later the regulation was overturned, and the couple quickly wed.

This personal happiness aside, the great tragedy of János's life was that his work was completely ignored. His tiny *Science of Absolute Space* was included almost a century later on a list of the ten most important mathematical books of all time. This may have been some gratifying comfort to the intellectual descendants of János, but it is a pity such recognition could not have come in his lifetime. *The Science of Absolute Space* was an absolute disaster from that point of view. With the exception of Gauss, nobody took notice at all.

János started other ambitious projects in mathematics, but he never finished them. In fact, he did not publish anything during the rest of his life. He amassed thousands of pages of notes that he never published and that nobody read while he was alive. His was the worst sort of tragedy. He discovered that if the fifth postulate were not true, geometry was still perfectly logical. His non-Euclidean geometry was strange, interesting, and important. But nobody paid any attention. Few knew what he had written and even fewer understood. János had climbed the mountain and discovered that he could fly. And then he climbed back down the mountain only to find that nobody cared.

A good example of this was when a mathematician named Antal Vállas wrote an article surveying Hungarian mathematics in 1836. He was a member of the Hungarian Academy of Sciences and well positioned to summarize all the important mathematical discoveries in his land. In his article, he mentioned Farkas Bolyai as someone who stood out as an eccentric but important mathematician. Even though he mentioned the *Tentamen*, nowhere did he note Farkas's son or his appendix to the book.

That same year, Farkas wrote to Gauss in 1836 about the dismal state of mathematics in his world. "Here nobody needs mathematics," he said. "Out of my students there are only a few who have a sense for it; I use my work as waste paper for wrapping up things and the like." Farkas added that his mathematical work had been of great use to him during the cholera epidemic. He had been infected and suffered from the bouts of watery diarrhea that are characteristic of the disease. He implied to Gauss that he used his life's work during this time as toilet paper to help himself get through the episodes of diarrhea.

•   •   •

Nikolai Lobachevsky married a year after Gauss's wife died and just before János Bolyai met his own wife. Varvara Alexeyavna Moiseyeva was the young daughter of one of the wealthiest families in Kazan, and when she married Lobachevsky in 1832, she owned a number of different estates in Russia and a huge house in Kazan. Lobachevsky and his wife filled their home with children and friends. They employed the best cooks, had the best parties, and enjoyed a lifestyle befitting one of the city's most fabulous and successful couples.

These parties were just a facade, however. Behind the veneer, Lobachevskey and his wife were in reality a very unhappy couple. They did not have a lot in common and were so incompatible that they often fell into long and bitter arguments. Sometimes Varvara would berate Lobachevsky brutally while he silently paced the room smoking his long pipe. His life had become an absurd comedy.

Like János Bolyai, Lobachevsky had also stared down a bottomless chasm. Like Bolyai, Lobachevsky had jumped. He soared. He returned. And like Bolyai, he still couldn't get anyone to care about what he had done. He continued to work on his "imaginary geometry" and wrote a series of memoirs on the subject from 1835 to 1838. He even adapted his work for the rest of Europe and published a short treatment for a journal in 1837. None of this did any good.

Lobachevsky was a lonely math scholar from a remote part of eastern Europe—a diamond in an intellectually poor Russian rough. On top of that, he was a very bad writer. In Russia, he was read little and understood even less. Where Lobachevsky's great invention was not ridiculed, it was ignored.

14

The Birth of Electronic Communication

I don't remember my having made any previous mention to you of an astonishing piece of mechanism that we have devised.
—Carl Gauss, letter to Heinrich Olbers, 1833

The Sunday churchgoers in the town of Göttingen discovered a new feature adorning the steeple of their beloved St. John's Church one week near the end of 1833. Two lines of curved wires, one up to the top of the church and one down again on the other side, had been installed.

Our modern urban landscape is so crisscrossed with electric lines of all types stretching from pole to roof to roof to pole that it seems strange to consider a single eight-thousand-foot stretch of parallel wires being worthy of mention, but one wonders what the townspeople thought. They had no telephones, no radio, no form of electronic communication at all. So they had no way of knowing that this was the means for the first successful electronic transmission of information ever.

The device was a galvanic circuit conducted through wires that were stretched from Carl Gauss's observatory up to the top of the church steeple and then down to the physics laboratory that his younger colleague William Weber managed. At both ends the wires were connected to coils surrounding heavy magnets, with Gauss at one end and Weber at the other.

"Carefully operating my voltaic pile, I can cause so violent a motion of the needle in the laboratory to take place that it strikes a bell, the sound of which is audible in the adjoining room," Gauss wrote. "We have already made use of this apparatus for telegraphic experiments, which have resulted successfully in the transmission of entire words and small phrases. This method of telegraphing has the advantage of being quite independent of either daytime or weather; the ones who receive it remain in their rooms, and if they desire it, with the shutters drawn."

Gauss saw the potential for this method of communicating, which was something that he grew to appreciate. The first time he mentioned the telegraph to anyone in a letter, he said that it served merely as an amusement. But then about a year later, he published a more detailed report about his early telegraph, which he called his great galvanic circuit. "The employment of sufficiently stout wires, I feel convinced, would enable us to telegraph with but a single tap from Göttingen to Hanover, or from Hanover to Bremen."

Solely by constructing a circuit using the standard electric and magnetic equipment available to them, Gauss and Weber made the first working electronic telegraph and were able to connect their two laboratories across a distance of five thousand feet. The first transmission was a simple summons intended for the servant Michelmann, who ran errands for the two.

The device came to be known as the Gauss-Weber current-reversing great galvanic circuit, and it represented a major step forward in the history of communications. Just a few years earlier, when Napoleon's armies were dominating warfare in Europe, the best form of battlefield communication had been large towers erected at appropriate intervals that allowed individual soldier-communicators in the towers to convey messages from one to another by using hooded lamps or some other means of purely visual communication. A message could be relayed from one tower to the next and so on. Technologically, speaking, these towers were barely more advanced than smoke signals.

Visitors marveled at the circuit's operation. The Duke of Cambridge visited the Göttingen laboratories and is said to have taken a special interest in the device. Some of Gauss's friends discouraged him from pursuing this endeavor further, however, calling it frivolous and unscientific. But Gauss saw the potential for the invention. He conceived of operating a telegraph alongside the train tracks. "Could [sufficient funds] be spent on it, I believe electromagnetic telegraphy could be brought to a state of perfection and made to assume such proportions as almost to startle the imagination,'' he wrote to a former student in 1835.

The previous decade had witnessed amazing advances in the field of electricity and magnetism, and several scientists in Europe were revolutionizing this area of physics. Perhaps the greatest of these was Michael Faraday, who had just discovered induced current, the ability of magnets to induce an electric current in conductors, and was beginning to publish just before Gauss constructed his great galvanic circuit.

Gauss was intensely drawn to the study of magnetism. This was a very old field, but in Europe the most exciting work was just now being done. Scientists whose names were later associated with the fundamental units of electricity and magnetism were then in their prime of discovery. Gauss was aware of the work of men like André-Marie Ampère, Michael Faraday, Hans Christian Ørsted, and Jean-Baptiste Biot. He was especially keen to get into this field as well. It was something of a new lease on life for him, scientifically speaking, coming shortly after Minna's death.

Gauss's interest was pushed forward when a twenty-four-year-old scholar named William Weber arrived at Göttingen to be the new professor of physics. The death of a professor had created a vacancy at the university, and the government bureaucrat in charge of filling it wisely asked Gauss whom he should hire. Gauss enthusiastically supported Weber for the position and even wrote him the letter of congratulations when he was selected. Soon, Gauss and Weber became fast friends and were often dinner guests in each other's homes. Though Weber was a generation younger than Gauss, his arrival sparked the final period of great productivity in Gauss's career.

Gauss had met Weber four years earlier at a conference in Berlin. They quickly formed a collaboration, did experiments together, and started a journal to publish their results. These were exciting times for Gauss. He really had very little to gain from a collaboration in mathematics. There were none in his day and very few in the history of mathematics who could have taught him much of anything at that point in his life. But Gauss was already interested in physics and had published works on the calculus of variations applied to physical problems related to electricity and magnetism. Magnetism was already an important and growing area in those days. Gauss was willing to work with someone who was bright and motivated, so he gladly welcomed the collaboration with Weber.

They formed one of the great research teams in history. Gauss was older and wiser, and Weber, more nimble in his youth, was particularly adept at instrumentation. Together they shared two brains, four hands, and one mind. Although Gauss was most successful for his theoretical work, it is a mistake to think of him as solely a theoretician. He had, in fact, a highly successful later career as a laboratory physicist, and it was through his collaboration with Weber that his experimental work bore some of its greatest fruit. From 1832 to 1834 the pair constructed their great galvanic circuit, a truly advanced bit of communicative wizardry.

Gauss and Weber had a plan to transfer their technology to the German railroad authority, which was then constructing a network of rails to link the urban centers of the land, although nothing came of it. Gauss thought that the same network could be a natural infrastructure to connect German cities via a telegraph. He envisioned burying copper wires alongside the rails that would connect to his apparatus in different cities and be used to carry messages. Weber was even more ambitious. He envisioned conducting current along one rail and back along the other so that no additional wires would be needed. In the end, the railroad authorities rejected both proposals because of the uncertainties and costs involved.

Still, this was a great discovery in the history of communications. Gauss and Weber were the first to use electrical currents to carry messages over long distances, though these distances are nothing compared to the separations over which electronic communication takes place today. The Gauss and Weber network was archaic even by later nineteenth-century standards. All it really consisted of was a few lengths of wires that connected two nearby points in Göttingen. A cannon fired at one of these two points— either end of the network—would probably have been clearly audible at the other. So in terms of practical innovations, Gauss and Weber did not improve much over what one might have accomplished with purely visible or auditory signals. But technologically, their invention was fundamentally more advanced than anything the world had ever seen.

Although Gauss and Weber made efforts to communicate their success, it was soon forgotten. They lived in a small city. The railroad itself did not arrive in Göttingen until 1854, and by the time it did, Gauss was on his deathbed, his great discovery long since forgotten, and the technology he first invented was being pursued avidly by people who never knew what he had done.

So obscure was the discovery that when the telegraph was taking off as a technology in the 1850s, the British historian Sir David Brewster was surprised to learn of Gauss's discovery. He wrote to the then aged Gauss at the end of 1854, "As I am, at present, writing on the subject, I would esteem it as a particular favor if you would oblige me by a notice of what you have done, and of the time when you used it publicly." Gauss's reply to Brewster was the last letter he ever wrote. Many years after he died, his great invention appeared at the World's Fair in Vienna in 1873 and again at the Chicago World's Fair of 1893.

Inspired by the collaboration with Weber, Gauss also attacked a different problem. In 1832 he began investigating Earth's magnetic field. At the time, scientists all over the world were making measurements of the field. In addition, numerous explorers and travelers were taking magnetic measurements all over the world, and one British colonel compiled hundreds of them. All of this data was feeding the possibility of a general theory of terrestrial magnetism.

Working out the theory of Earth's magnetic field appealed immensely to Gauss, and he toiled away on this problem for the rest of the 1830s. His interest in this subject went back years. In 1806, he already knew the basis for the magnetic field and could have published on it then, but he waited more than thirty years until he had made the measurements himself and could use those of others to back up his model.

Gauss embraced this new field with enthusiasm, and it grew around him. Through his efforts, he formed an association devoted to magnetism and started a journal on the subject. Gauss published more than a dozen articles in his own journal. He also commissioned an expensive and carefully constructed magnetic observatory connected to his astronomical observatory. In order to avoid ferromagnetic effects interfering with his instruments, he had the building constructed out of only copper, wood, stone, and mortar. There was no iron in the building to interfere with magnetic measurements. Even the nails were copper. For the next four years, starting in 1832, he made his measurements.

Gauss's magnetic measurements were so impressive that the famed Scottish physicist James Clerk Maxwell said the work should be a model for anyone engaged in measuring the forces of nature. His building and his experiments became well-known in Germany, and many visitors came calling.

Gauss published a major memoir on terrestrial magnetism in 1833, the *Intensitas vis Magneticae Terrestris ad Mensuram Absolutam Revocata* (The Intensity of the Earth's Magnetic Force Reduced to Absolute Measurement). In this work, later called one of the most important papers of the century, Gauss reduced magnetic intensity measurements to units of mass, length, and time. He also introduced the science of variations in Earth's magnetic field, of which not much was then known, and derived the mathematical theories relevant to its study. Gauss theorized that the cause of Earth's magnetic field was the presence of polarized iron at the planet's center and near its surface, and he designed ways to measure the field strength at Earth's surface.

Gauss made the first absolute measurement of Earth's magnetic field. He suspended a magnet and studied its oscillations. He showed how to determine terrestrial magnetism in terms of the intensity of the field, as measured by the inclination and declination of a magnetic needle. He introduced the units of magnetic flux density. In his honor, by international convention one standardized measure of magnetic strength today is called a gauss. (Weber also has a unit named after him.) A magnetic field of one gauss exerts a force on a current-carrying conductor placed in the field equal to 0.1 dyne per ampere of current per centimeter of conductor. In more familiar terms, Earth's magnetic field, as measured at its surface, is about 0.5 gauss, a small iron magnet is about 100 gauss, and a large electromagnet in a modern magnetic resonance imaging (MRI) instrument may be 30,000 to 40,000 gauss.

Gauss was suddenly almost happy again. "I occupy myself now with the Earth's magnetism, particularly with an absolute determination of its intensity," he wrote to his friend Heinrich Olbers on February 12, 1832, just a few months after Minna passed away. "Friend Weber conducts the experiments on my instructions."

By the end of the decade, Gauss had published an atlas of measurements dealing with Earth's magnetism and calculating the location of the magnetic south pole extremely accurately. In 1841, an American sailor named Captain Charles Wilkes reached Earth's magnetic south pole and discovered it to be almost exactly where Gauss had predicted. Another navigator had found the magnetic north pole close to where Gauss had predicted it would be.

The wires that formed Gauss and Weber's great galvanic circuit remained there for more than a decade, until they were destroyed in an electrical storm. Lightning struck the wires on top of St. John's steeple on December 16, 1845, and blasted them into small pieces a few inches long and into tiny specks of metallic flakes. "All of which formed a brilliant rain of fire," recounted Gauss at the time. Luckily, no one was hurt. The circuit itself outlived the collaboration of its two inventors, Gauss and Weber, which ended much sooner than it should have when Weber and six other professors (infamously known as the Göttingen Seven) were expelled from the university for political reasons.

The years toward the end of Gauss's career were a time of great social change in Europe. In 1830, the French deposed Charles X, and this sparked similar protest movements across Europe, including in Germany. Gauss didn't take these things seriously, but many of his contemporaries certainly did. Revolutions fulminated all over Europe, usually to be swept aside in most cases by the more established power of existing governments. Universities were becoming more political in the nineteenth century, and nowhere was this more evident than in Göttingen.

The University of Göttingen had been emerging as a center of popular protest. It was celebrating its centenary in 1837, which included the construction of new buildings, parties, a torchlight parade, celebratory lectures and poems, and a massive dance party with thousands of people. To professors at the university, Göttingen must have seemed like the center of the world.

In Germany after the Napoleonic Wars, there rose a true middle class. While Gauss was the direct beneficiary of this, he was also a throwback to an earlier, less self-deterministic time. He was part of the last wave of a long line of scholars and other intellectuals who came to their stations through the generosity of a noble benefactor.

This may have played a significant role in his decision to steer clear of the political pit into which Weber fell.

The complicated crisis arose because King William IV of England, whose dominion included the Hanoverian state in Germany, granted a new constitution in 1831, partly in response to a popular uprising and student protests that took place in Göttingen in the previous months. The new constitution was more democratic and liberal than before and was met with a wave of popular acceptance. The crisis seemed to have been averted.

A few years later, a separate legal question created a new constitutional crisis. When William IV died without an heir in 1837, he was succeeded by his niece Victoria, who sat on the English throne for many years. By consequence of an obscure ancient law, however, Victoria could not (as a woman) simultaneously be queen of Hanover. So instead her uncle Ernest Augustus, the Duke of Cumberland, became king of Hanover.

The king was a brutal ruler, unsympathetic to the common man, and unmoved by popular protests except to become even more iron-willed. Hanover's new king wasted no time in wading into controversy. One of his first acts on the job was to use a legal loophole to declare the brand-new constitution null and void. This reactionary stance was motivated by a strange set of circumstances, largely personal. The king voided the constitution because it stipulated, in the most "enlightened" and unyielding terms, that any future heir to the throne would have to be free of physical defects. The king's son was blind.

Seven professors stood in defiance of the nullification and declared that their oaths to the 1831 constitution came before their loyalty to him. They registered their displeasure in a letter of protest sent directly to the king. The Göttingen Seven included the brothers Grimm and two people who were very close to Gauss—his son-in-law, Ewald, and his close collaborator, Weber.

To the king, these protests were no more significant than the bleating of sheep. He declared that he could hire university professors as easily as ballet dancers. The Göttingen Seven were summarily fired, and though they advocated on their own behalf and enjoyed popular support throughout Germany, they would not immediately be reinstated. They left the university a fractured place with a seriously damaged reputation that would not recover until the end of the nineteenth century. The loss left Gauss shattered as well.

While Gauss hated Napoleon and the French dominance of Europe, he never in his life showed any real interest in politics. He would not have been interested in joining the Göttingen Seven for this reason alone. But on top of that, he was a relatively old man at the time and was caring for his ninety-year-old mother. She was blind, and he was in no position to risk everything for the sake of a political statement, even if he had been inclined to take a public stance on a matter of politics, which he was not. But now, even though he was an indifferent, uninvolved bystander in this matter, it affected him deeply.

In losing Ewald, Gauss was also losing his daughter Minna, who was married to Ewald. And in losing Weber, he was losing the greatest collaborator he had ever had. These losses were severe for Gauss—both devastating blows. He wanted to do anything he could to stop them, and he lobbied heavily on Weber's behalf, approaching the local officials and pleading the case that Weber was merely a casual, incidental signer of the document, not one of the main conspirators. When those efforts failed, he wrote to his friend Alexander von Humboldt, who would have the opportunity to meet the king. He pleaded with von Humboldt that his entire scientific endeavor was in jeopardy because of the loss of Weber.

Humboldt, however, could not broach the subject with the king. Instead he presented Gauss's case to two high court officials. These men, however, knew that the king was profoundly displeased with the Göttingen Seven. The king was not about to be moved by the pleadings of one mathematician, however brilliant or well connected. The king's disposition would remain hostile, and so his courtiers would be unsympathetic. The effort failed. In 1838, Weber left the university. Thus ended the best collaboration Gauss had enjoyed in his life.

Once Weber and Gauss were separated, their partnership more or less ceased to exist. Years later, in 1849, Weber returned to Göttingen, taking up his old post at the university, but by then Gauss was in the last years of his life and more or less retired.

Perhaps the greater blow for Gauss was losing Ewald, a colleague and friend, not to mention his son-in-law. Ewald took Minna, his youngest surviving child from his first marriage, and his favorite, with him in exile to the German city of Tübingen. Gauss never saw her again. The summer after their departure, Gauss suffered from temporary deafness with loss of memory, vision, and teeth. This lasted for several months and disappeared after he tried a number of home remedies, including almond oil, which he believed was responsible for abating the illness.

The following year, on April 19, 1839, Gauss's mother died at the age of ninety-seven, having spent the last twenty-two years of her life with her son and the last four of those blindly feeling her way around the observatory. She lived out her last days confined to her room and completely dependent on her son. Though Gauss's mother had lived long, she had experienced a hard and brutal life—one Gauss himself described as being full of thorns.

This was not a very happy time for Gauss, either. In the summer of 1840, the year after his mother died, his daughter Minna passed away. After he heard the news, he wrote to Ewald that he was beside himself with grief. "Even now I cannot realize that my darling angelic child is lost to us on this Earth," he wrote. "It was always my dearest most consoling hope to be reunited with her here, and to see my last years thereby cheered. Now it is gone, this hope!" He once said of Minna, "The earth rarely sees such absolutely pure, noble creatures." That same year, Olbers, one of Gauss's oldest friends, also passed away.

So the 1830s, a decade that had begun with such promise for Gauss, was to end in misery. Perhaps the only bright spot was in November 1838 when the Royal Society of London gave him the Copley medal, the society's highest honor. Characteristic of his mood, Gauss wrote to his daughter that if the metal of which the award was made had been worth more, he would have sold it and given the money to his children.

Non-Euclidean geometry was probably the furthest thing from his mind in the beginning of the 1840s. He began spending time on séances, which were then becoming very popular. But he was about to rediscover non-Euclidean geometry when the work of an obscure Russian scholar became available in Germany.

15

The Imaginary Man from Kazan

What Vesalius was to Galen, what Copernicus was to Ptolemy, that was Lobachevsky to Euclid.

—William Kingdon Clifford

The more Nikolai Lobachevsky worked on his imaginary geometry, the more he became convinced that it was a true form of geometry. He knew what he had created, and he was convinced that the day would come when other mathematicians would come to appreciate what he had done. They only needed to be aware of and get access to it.

If we could observe more of the universe, Lobachevsky believed, then we might just be able to observe his imaginary geometry. It might even be obvious to us. He thought that since we inhabit a small corner of the universe, non-Euclidean geometry seems foreign. He wanted to do everything he could to make it seem less so.

Lobachevsky was a visionary in this sense. He invented new words such as *horocycle* and *horosphere* to go along with his new form of geometry. "A completely new science was created, new ideas and facts were ushered into the world, which bore the genius of their maker," one of his biographers wrote. "Human knowledge had never known anything like it."

From 1835 to 1838, Lobachevsky published his *New Elements of Geometry* in the obscure scientific journal of Kazan University. This time, he made no attempt to garner the notice of Ostrogradskii and the St. Petersburg Academy. Instead, he turned farther west. Publishing his work in the rest of Europe made a lot of sense to

Lobachevsky, since he was unable to get anyone in his native Russia to take notice—neither in Kazan nor in the more metropolitan cities like St. Petersburg. In 1840, Lobachevsky's "imaginary geometry" was finally translated into French and German and published in Berlin. Unfortunately, Lobachevsky was a poor writer, and his manuscripts were hard to penetrate. If his work was short and confusing to begin with, the problems were compounded because the translations were very bad, making his work almost unintelligible. Carl Gauss, who eventually read Lobachevsky, compared the work to "a confused forest through which it is difficult to find a passage and perspective, without having first gotten acquainted with all the trees individually."

Ironically, Gauss first heard of Lobachevsky when he read an unfavorable review of his work in a German periodical. Gauss was himself an incredibly clear writer who had a masterful command of his German language and wrote very elegantly. He was quite talented with language in general and might have followed a career as a linguist of sorts had he not been so well suited to mathematics.

There was enough in what Gauss read for him to recognize something in Lobachevsky. After all, he had puzzled over the same problems enough times to know when someone had come to the same conclusions in geometry as he had. Lobachevsky had invented a form of non-Euclidean geometry sure enough. He called it "imaginary," but Gauss knew exactly what it was. He managed to cobble together a number of pieces of Lobachevsky's work in order to study them and confessed to a friend that he got exquisite enjoyment from reading them.

Gauss claimed that Lobachevsky had taken a path different from the one he had traveled, but it covered the same ground. He was so impressed that he paid Lobachevsky the ultimate compliment of learning Russian so that he could benefit from reading all of his writings in their original language. This is not as extraordinary as it might seem, however, because Gauss had an incredible facility for learning new languages—even late in life. For instance, he started reading Sanskrit when he was in his sixties.

Lobachevsky's work got no help from Gauss, however, because after Gauss read it, he did not go out of his way to promote it. He did write to his friend Heinrich Schumacher, saying that he wanted to draw his attention to the work, which he promised would give Schumacher "thoroughly exquisite pleasure." But he never published a review or anything else about his Russian contemporary. Instead, he sat idly by as more articles soon appeared in German publications deriding Lobachevsky's work.

Like the efforts of an Anton Chekhov character, Lobachevsky's effort to disseminate his non-Euclidean geometry was utterly futile. He knew he had invented something profoundly important, and he worked for years to perfect and demonstrate its mathematical significance. The tragedy of his life is that nobody listened. "Imaginary geometry" was an unfortunate term because it was almost an oxymoron. Geometry was real, not imaginary. Lobachevsky nevertheless believed that non-Euclidean geometry was the true form of geometry—a reflection of reality. He denounced mathematics that was not grounded in reality, in fact, saying, "All mathematical principles which are formed by our mind independently of the external world will remain useless."

In the end, Lobachevsky did profoundly influence mathematics. He was one of the first to incite a change in the field that would culminate, by the end of the century, with mathematicians regarding the laws of mathematics as arbitrary inventions rather than as the laws of physics or other sciences represented in mathematical terms. In his own day, there were very few people who could understand his work, though.

In 1846, Gauss wrote to a friend that Lobachevsky's work was presented with "masterly skill in the true geometrical spirit." But these words were never published while Gauss was alive. And when they finally were, the recognition came perhaps too late.

Though Lobachevsky never received the credit he deserved, his failed efforts to promote imaginary geometry did not cause his reputation to suffer. He remained a respected mathematician. Perhaps the people who knew him would joke about his imaginary geometry. Perhaps they would think it quaint or pitiable that he had his head in such imaginary clouds. Nobody ever seemed to lose esteem for him, though, and his career continued without interruption. In 1835, Moscow University elected him an honorary professor and lauded him as "one of our best professors of mathematics."

After Gauss read Lobachevsky's work, he recommended Lobachevsky to the Royal Society of Sciences in Göttingen, calling him one of the most distinguished mathematicians in Russia. Many of

Gauss's contemporaries would have considered Kazan (and perhaps the rest of Russia as well) to be a scientific backwater, but Gauss had long had a warm affection for and a deep connection to Russian mathematicians and scientists. He had been a corresponding member of the Academy of Sciences at St. Petersburg since his prediction of the orbit of Ceres. He had even been wooed by offers to lead the observatory at St. Petersburg. Upon such a strong recommendation by its most famous member, the academy elected Lobachevsky immediately.

Elections to academic societies were a nice honor and sometimes a sort of academic currency to be traded and owned. One individual who was influential in an academic society could help ensure the election of foreign and domestic members and would do so. This honor not only brought prestige to an individual but to the society itself—the more august its body of members, the more bragging rights a society would have. Lobachevsky was thrilled that the honor was given to him by no less a person than Gauss, whom he had admired since he was a child. He repaid Gauss by having his university give Gauss an honorary doctorate. Still, Lobachevsky was a sad, lonely genius who never once got to discuss his greatest invention with someone who listened and wondered enthusiastically at what he said.

In the end, Gauss avoided any potential controversy, and he never mentioned Lobachevsky in anything he published. He also never told Lobachevsky that János Bolyai had done the same thing, even though Gauss had already studied the latter's work as well. Nor did he tell his friend Farkas Bolyai that a Russian mathematician was publishing the same work as his son's. It was only a matter of time, though, before the Bolyais found out about Lobachevsky.

In the early 1840s, a Hungarian mathematician named Franz Mentovich went to Göttingen and met Gauss. There he learned for the first time of Lobachevsky's work in non-Euclidean geometry. Either Mentovich knew of János Bolyai's work or Gauss told him about this as well, because sometime later, in 1844, Mentovich made his way to Marosvásárhely, where Farkas Bolyai was living. Farkas's son, János, was living close by. When Mentovich met with Farkas, he told him about the work of this Russian scholar from Kazan, a city in the remote steppes on the border of Siberia. He said that Lobachevsky had invented a non-Euclidean geometry that was very similar to János Bolyai's and that Gauss knew all about it.

Farkas and János tried to get a copy of Lobachevsky's papers. Failing to do so, they wrote to Gauss asking if he could obtain a copy for them. Gauss wrote back sometime later telling them that Lobachevsky's work had been translated into German and that they could probably find one of these translations. On October 17, 1848, father and son found a single copy and read it.

When the Bolyais first saw Lobachevsky's work, they were excited and thought about writing some sort of review or reply. They next made numerous notes on the manuscript. János made a careful analysis of it, in particular studying Lobachevsky's work to see how it compared it to his own appendix. He found rich connections between what he had developed and what Lobachevsky had written.

Father and son planned to publish a response. But Farkas was an old man and János was overtired from disease and disappointment. So what might have been the beginning of a wonderful correspondence amounted to nothing. In the end, they gave up trying to work together on any sort of review of the work. Farkas did a nice review of the work on his own, in a small book published just before his death. But this did little to raise awareness of a subject that the world seemed to care nothing about.

János took a much darker view of the whole thing. He had become so disgusted with the lack of attention to his own work that he was verging on depression. In some paranoid fit, he seems to have formed the opinion that Lobachevsky was a fiction—a pen name created by Gauss to cover up the fact that he had stolen his ideas. He thought Gauss had come up with an elaborate scheme to take his work out from under him.

The stage by now was set for a full-on priority dispute between the three mathematicians. History is filled with such disputes. The discovery of calculus and the dispute between Isaac Newton and Gottfried Wilhelm Leibniz over which of them had done it first is a famous example of this. But between Gauss, Lobachevsky, and János Bolyai, no dispute ever erupted. Part of the reason was that when they were alive, there was not much at stake. Newton and Leibniz were both incited to fight for all the credit for inventing calculus because it was almost immediately embraced as a great mathematical invention.

This was not the case with non-Euclidean geometry. Bolyai and Loba-chevsky had both published their work, but nobody cared. Gauss avoided doing so because he thought it would be more trouble than it was worth.

Besides that, Newton and Leibniz hated each other when they were old and fighting over which of them deserved credit for calculus. János's paranoid resentment of Gauss aside, he, Gauss, and Lobachevsky did not hate one another—perhaps the opposite. As different as their lives had been, they had much in common. They had each scaled the same mountain that so many mathematicians before them had climbed, and they had each made the leap of faith that it took to invent non-Euclidean geometry. But they never broke into a fight over who deserved the credit. Instead, each man in turn shrank from his own discovery.

It was as if non-Euclidean geometry was a burden that each man was cursed to bear in his own way. For Gauss, the discovery brought to the surface a strange paranoia, and he recoiled from non-Euclidean geometry even though he clearly understood its importance. János and Lobachevsky completely failed to attract attention for their invention, which a century later would be regarded as one of the most profound advances of all time. János thought his discovery would bring him fame and fortune. Instead it brought him nothing but grief, and in the end he gave up caring about it altogether. For Lobachevsky, it brought nothing at all, and his works were ridiculed by the few people who read them.

So while the invention of non-Euclidean geometry should have been the sweet reward for two thousand years of failed efforts, for Gauss, Lobachevsky, and János Bolyai, it only meant knowing that they had discovered something great but failed to convince others of its significance. They all knew that they had never really been recognized for it. Thus what might have been a contentious three-way argument over which of them deserved credit fizzled before it erupted.

In fact, there has been more controversy in the twentieth century and today over which man deserved credit. Some have argued over whether Gauss should be considered an inventor at all, since he never published anything on the subject in his lifetime and left only a few scattered notes among the twelve thick volumes of collected works that were published after he died. Others have argued that János Bolyai should receive equal credit with Lobachevsky for having the first publication, since the imprimatur of Farkas Bolyai's *Tentamen* is 1829, the same year that Lobachevsky published his treatise in the *Kazan Messenger*.

On July 16, 1849, the city of Göttingen got behind its most famous resident and celebrated the fiftieth anniversary of his graduation. The city hosted a "Golden Jubilee" for Gauss, a high point at the end of his career. From near and far they came—scholars, friends, relatives, friends of friends, and friends of strangers. They brought warm feelings, gifts, honorary doctorates, titles, and other honorifics. From many more people who could not attend came letters and notes of congratulations. "There was no end of letters and communications," Gauss's daughter wrote to her brother about the time. Gauss was celebrated in speeches and written works from former students and colleagues throughout Germany and beyond. He received honors from several cities, and the king wrote him a note congratulating him. Numerous intellectuals and dignitaries came from far and wide to honor him. Flowers were strewn all over the city, and a great banquet was held.

One thing that they didn't discuss in all these speeches and letters was non-Euclidean geometry. Few of Gauss's acquaintances knew anything about it, and it was as if the subject did not exist. Still, the year 1849 was a definite high point in Gauss's later life. William Weber returned to Göttingen that year, and though Gauss was too old to renew their earlier fruitful collaboration, he was glad to have his friend back. For the next five years, Gauss would be winding down his career.

This was a time of great change in Germany. Europe had been largely rural at the end of the eighteenth century. Indentured servitude was still commonplace in eastern Europe, and everywhere political and economic life was dominated by nobles and landowners. This was changing as the continent was witnessing a season of liberalization. The population of Europe nearly doubled in the eighteenth century and more than doubled again in the next hundred years. In 1800 it stood at 205 million, and by 1900, there were 481 million people living in Europe.

Gauss, Lobachevsky, and János Bolyai would all feel its effects. But the revolution even closer to their hearts remained silent. Non-Euclidean geometry—their great revolution in mathematics—was to free geometry from the shape of our three-dimensional world and send mathematics into worlds of which Euclid never dreamed. But this revolution was nothing but a dream in 1849. For Gauss, Lobachevsky, and János, their greatest years of discovery were well behind them.

In 1854, the year before his death, Gauss took his last trip away from Göttingen when the new railroad opened. He traveled to Hanover overnight and back. The times were changing all over. The railroad was opening up the West in the United States just as it was connecting the old cities of Europe. The bumpy, dusty creaky buggy rides of old would fade into a steam-and-smoke reality of the new industrial age.

That same year, Gauss attended a lecture given by a young mathematician in his department named Bernhard Riemann. The lecture was one of the requirements of Riemann's new position. Gauss was not required to attend, but he was excited to do so. On the way home from the lecture, he filled Weber's ears with praise of the young mathematician and his brilliant ideas.

Gauss had had a good life, and now was the time to appreciate it. He raised five children, though one had died and two moved to the United States. He had eighteen grandchildren, most of them born while he was still alive. He met only one of these grandchildren in person, however, since the rest were born in the United States. He buried two wives and his mother, who had lived with him for the last two decades of her life. But in the balance, it was a good life. In his final years, he followed a very regular schedule, going to his club, reading newspapers, and occasionally playing cards with his friends.

All three inventors of non-Euclidean geometry died within a few years of one another, starting with Gauss in 1855. Lobachevsky died the following year, and János Bolyai passed away in 1860.

When Lobachevsky died on February 26, 1856, he was a blind, lonely, broke, abandoned, and forgotten man. In 1846, the same year that János Bolyai was reading the work of his Russian contemporary, Lobachevsky had to vacate his chair as professor, which he had held since 1816. The university bylaws required that no professor should serve more than thirty years in his chair, and his term was up. Even though a petition was circulated to keep him on as rector,

Lobachevsky decided to resign so that his younger protégé could obtain the post.

He had enjoyed a rich career as a scholar and was a most popular rector at Kazan University. On the one hand, he had risen quickly through the ranks, becoming the top mathematician and leader of the university—one of the best. He saw education as something that should destroy ignorance and enhance moral character and make his students physically and mentally healthy. He dabbled in architecture and helped to build an observatory, an operating theater, and other buildings. He was reappointed to this position six times after standing for a vote by the faculty. He turned out to be a smart and a tireless administrator.

He became assistant guardian of the Kazan district and served in this role until 1855. That same year, the university celebrated its fiftieth anniversary and Lobachevsky published his last paper. *Pangeometry* was not a new work but rather one final exposition of his non-Euclidean geometry. He was blind by then and had to dictate much of this work.

The last few years of Lobachevsky's life were tough on him. He retired in 1846, and his health rapidly deteriorated. His oldest son died, and in the end, the world did not care about imaginary geometry.

When Lobachevsky married Lady Varvara Alexeyavna Moiseyeva in 1832, she was a young girl from a wealthy family with a sizable fortune of her own. Marrying her had helped improve his social standing, but it was a disastrous marriage, both emotionally and financially. Despite his prominence and his wife's generational wealth, the finances of the Lobachevskys suffered deeply from their excesses. Part of the problem was that Lobachevsky sold off three of his wife's estates in order to buy one in the countryside near Kazan.

Lobachevsky loved being a gentleman farmer, and it was his goal to convert his new estate into a model farm. He built a house on the land and added a wing to the house. He built a mill, a barn, and a threshing house, and he planted an orchard. He dammed a creek and paid countless workers in the process. All the while, he lectured on livestock feeding, crop rotation, watermill construction, and the proper storage of potatoes through the winter. The one thing he didn't lecture on, apparently, was proper management of the farm budget—a subject for which he was ill prepared.

He was a spendthrift and liked to live large, gambling and attending plays. His expenses were considerable, well in excess of his income, and his wife had to help bail him out by mortgaging her house in Kazan. Later they had to borrow money, and after Lobachevsky died the whole estate had to be sold to settle the outstanding balance.

There were other miseries as well. They raised seven children and had several more who died at an early age. Just after Lobachevsky retired, his favorite son, Alexi, died of tuberculosis. Then there were the professional failings. In the last few years of his life, Lobachevsky was working on completing his life's work on geometry. He had never stopped trying to publish and raise awareness of his invention of non-Euclidean geometry and his other work, though by the 1850s he had been more or less forgotten by his colleagues, and his ideas were set to be ignored.

Lobachevsky died depressed, confused, and utterly ignored. His sad, tragic life reads like a story by his fellow Kazan University alumnus Leo Tolstoy. Or perhaps more aptly he was like a Chekhov character—feeble, blind, fierce, possessing of great ambition, and utterly destroyed by the nineteenth-century Russian culture in which he lived.

Lobachevsky was not fully appreciated until more than a decade after his death. But by the end of the century, he was honored in a way befitting of his genius. One hundred years after his birth, a large celebration was held in his honor. In 1893, in commemoration of his centenary, the university faculty met with distinguished foreign professsors in Kazan and held a day of addresses commemorating his life and work. A commemorative plaque was placed on the front wall of the house where he lived. In 1909, Lobachevsky's early geometry manuscript from 1823 was rediscovered and printed.

By the middle of the twentieth century, Lobachevsky had become an even more legendary figure, and during the rise of the Soviet Union he was something of a state hero. His stubbornness, reported atheism, and genius supported his rise as a champion of the proletariat. To the Soviets, Lobachevsky represented not just the greatness of the common man, emerging from a humble background as he did, he also was a revolutionary of sorts. Bold in the face of thousands of years of tradition, he was willing to throw off the accepted paradigm and strike forth into this strange new world of non-Euclidean geometry.

• • •

Farkas Bolyai died the same year as Lobachevsky, on November 20, 1856. His son János survived him by a few years. János had moved away from his family in 1852 and lived by himself with a servant who cared for him, as he was in poor health. He was despondent and on the verge of insanity in his last days. He was tortured, no doubt, by years of chronic infections and poor health, the lingering effects of having suffered a traumatic head injury a few decades before, and perhaps even the great disappointment of having to take with him to the grave the knowledge that he had solved the two-thousand-year-old mystery of the fifth postulate and nobody recognized him for it. This time, on his way to eternal rest, there was no climbing back down the mountain.

When János finally died on January 27, 1860, his servant sent a short note to his family. "The captain is gone," the servant wrote. János was given a burial with a military escort. Someone at the church where he was buried contributed an anonymous entry into the church registry that recorded János's funeral. The anonymous note celebrated the forgotten genius. "He was a famous mathematician of great mind. He was first even among the first. It is a pity that his talent was buried unused."

It is fitting in a way that the only voice appreciating János Bolyai at his death was an anonymous one. He never received the recognition he expected when he was alive, so why should he have expected anything else at his death? He died a complete unknown— a forgotten genius who had never actually been discovered.

Nine years after János Bolyai died, the Hungarian minister of culture, Baron József Eötvös, who was president of the Hungarian Academy of Sciences, received a letter from a Prince Baldassare Boncompagni in Italy. Prince Boncompagni was pleased to inform Eötvös that the biographies of Farkas and János Bolyai had just been published in Italian and that they would soon be sent, along with the Italian translation of János's *Science of Absolute Space.*

The mathematicians in Rome, Prince Boncompagni said, considered János Bolyai's work to be among the greatest mathematical achievements of the century. In fact, he seemed to be more famous elsewhere in Europe than in Hungary itself. The French mathematician Guillaume-Jules Hoüel once even wrote, "I am grieved to see how little Hungary appreciates her own scientific results."

Baron Eötvös was floored by this letter. He may have had little or no inkling that his own countryman János Bolyai had accomplished anything, let alone a work for which he was recognized by some of the best mathematicians in János Europe. Because Baron Eötvös was the Hungarian minister of culture, this was terrific news. It brought great honor to his country. And because he was the president of the Hungarian Academy of Sciences, this was also great news, because the honor belonged to a Hungarian who was rightly a part of that tradition.

Baron Eötvös began to champion his forgotten countryman. Within a few years, the Hungarian Academy of Sciences established a prestigious prize in János Bolyai's name. Eötvös arranged for all of János's notes to be archived, and what an archive it was! When János died, he had only the one publication to his name, but he left some fourteen thousand pages of mathematical notes and manuscripts. Early on, János's letter to his father from the 1820s, half a century before, was discovered among these notes. The János who wrote this letter was a young, dashing army man, a brilliant swordsman, a sharp soldier, and a raging genius—decades before disease, failure, and misery tore him down. Like a voice beyond the grave, here was János announcing to his father his incredible discovery of non-Euclidean geometry: "From nothing I have created a strange new world," he wrote.

This profound and personal note became a famous reference among mathematicians. As appreciation for non-Euclidean geometry grew, János's fame grew as well—both in Hungary and elsewhere. By the end of the century, his work was published in Hungarian and a half dozen other languages. Long before, in 1832, it had appeared in Latin alone.

In the twentieth century, János Bolyai's genius emerged in full flower. Libraries and universities were named after him. Hungary issued a commemorative stamp with his likeness to celebrate the hundredth anniversary of his death. Finally the recognition he knew he deserved had arrived—a century late.

Gauss was already a national icon when he died at the age of seventy-eight. He has been recognized as one of the greatest geniuses in mathematical history, and he is certainly regarded as one of the most important German thinkers of all time. In the twentieth century his likeness adorned the ten–Deutche Mark note.

Gauss began to grow more ill toward the end of 1854, and he was finally bedridden. He had always been very close to his daughter Theresa, but in his final days he demanded that she stick by his side almost continuously. She did this, but it seems to have been a slightly uneasy task. She claimed that Gauss even said to her, just days before he died, that he would have preferred them to die together. "The best and greatest that God could grant us would be this one favor, that we two on the same day might die together," Gauss said, according to Theresa. He died in the middle of the night on February 23, 1855.

Theresa Gauss, freed from all responsibilities of taking care of her father, could now pursue her own interests. Those included marrying the man who had been her romantic interest for several years—Constantine Staufenau, an actor and theatrical producer. The match raised a few eyebrows.

Gauss was as shrewd with money as he was with math, and he socked away a fortune in his lifetime. He had a massive amount of cash secreted in his desk and dresser drawers and other places. This was not the only presumed treasure that he left behind. His brain was regarded as a one-in-a-million specimen to be saved and studied. Rather than consign it to the anonymous erosion of posthumous rot, the pathology department at the University of Göttingen snatched it up, preserved it, and measured Gauss's skull carefully. Embalmed, the brain stayed in deep storage for some future generation that understood these things better to probe. What was the secret of his genius?

Gauss's brain weighed about 3.3 pounds and had a large cerebral area. One report said that Gauss's brain had deeper than normal convolutions, and more of them as well. The deep and frequent convolutions were speculated to be part of the secret of his genius. This was strictly anecdotal, of course. A recent, much more scientific study by neuroscientists using magnetic resonance imaging in the late 1990s probed deeper into the brain. Scientists found nothing unusual about its gross anatomical features, which is not to say that he had an ordinary mind—just one that stymies attempts to identify its unique features after more than a hundred years floating in embalming fluid.

More valuable than Gauss's brain was the legacy of books and notes he left behind. He carried much of his work with him to the grave, but he left behind a rich treasure of papers, letters, and various notes. It took scholars nearly seventy-five years to edit his collected works.

It was only after Gauss died, when people started going through his notes and files, that interest in non-Euclidean geometry became more pronounced in Europe. Other mathematicians began to discover what Gauss seemed to know, and they too became interested in non-Euclidean geometry. Gauss was a very famous man, and his materials related to non-Euclidean geometry caused many to take notice.

Before Gauss, Lobachevsky, and János Bolyai died, nobody knew what they had accomplished by inventing non-Euclidean geometry. Lobachevsky had endeavored to print his ideas, but his work was so opaque that he failed to garner any attention. Bolyai published only one work in his life and suffered from such a chronic failure to get his ideas accepted that he wound up turning away from mathematics in disgust. It's not exactly clear why Gauss never published, but the likely reason that he did nothing in the 1810s and 1820s was that he preferred to avoid the controversy that he knew such publication would evoke, and besides, he had much more pressing and important work as he saw it. Perhaps he even thought that once he passed away, people would see his notes on the matter and form their own opinions, which is exactly what happened. In the 1830s, surrounded by death, when he finally decided to publish some of his thoughts on non-Euclidean geometry, the likely reason that he still did nothing was that he saw János Bolyai's published appendix. With that, Gauss thought he no longer needed to publish.

Passages in books Gauss owned that discussed the fifth postulate were scrutinized for notes he may have made in the margins to see what he was thinking about the subject. The ground began to be prepared for understanding and ultimately accepting non-Euclidean geometry. Other mathematicians took up the challenge and started to study the work of Gauss, Lobachevski, and János Bolyai. Non-Euclidean geometry was growing in the second half of the nineteenth century.

Non-Euclidean geometry was a profound and far-reaching development that would change the nature of mathematics.

It transformed mathematics from a set of tools for dealing with real magnitudes to those dealing with merely with relationships—whether real or abstract. To accept non-Euclidean geometry required not only engaging in abstract thinking, but also rejecting the material world—staring out into the abyss and jumping.

In this sense, the most important thing the subject did for mathematics was to free geometry from what we understand to be external reality—the shape of our three-dimensional world. It sent mathematics into worlds that Euclid never dreamed of, where geometries were logical, consistent, and practical, but not necessarily Euclidean. Essentially what it did was to say that reality and geometry were different. Instead of representing real lines in the real world, non-Euclidean geometry set imaginary lines in a strange new world. It created a whole new way of looking at the shape of space—one where instead of only one line was parallel to another through a given point, there were infinitely many (or none at all). The interesting thing about non-Euclidean geometry was not whether it was a true representation of the world but whether it worked.

All these concepts were hard to swallow, though, because geometry had always been a natural science. From measuring plots of flooded land in Egypt to plotting the orbits of the planets, the lines and shapes described by geometry always corresponded to the objective real lines and shapes and of our world. Thus many mathematicians in the late nineteenth century were resistant to the notions of non-Euclidean geometry. Even as it became known, it was rejected by many who recognized that traditional geometry had described reality perfectly for more than two thousand years. Up to this point, only one non-Euclidean geometry had been found—the assumption that the fifth postulate was not true. It would only take a few decades to discover the next type of geometry.

Ironically, it was only after Gauss, Lobachevsky, and János Bolyai were all dead and their efforts to solve the mystery of the fifth postulate were decades in the past that the real growth of non-Euclidean geometry occurred.

16

The Soul of the Universe

Every great advance in science, every great discovery in nature, and every great invention has had its crowd of ridiculers; and non-Euclidean Geometry is no exception.

—*American Mathematical Monthly* editorial, 1895

A few years before Gauss died, he crossed paths with the second most outstanding German mathematician of the nineteenth century. Bernhard Riemann was just beginning his career at Göttingen as Gauss was ending his. They overlapped by just a few years, but their brief interaction was one of the greatest times in the history of geometry.

The two men had no collaboration to speak of. They produced no work together. Gauss was well past his productive years, and he was at a time in his life when his routine was to sit and read the newspaper for hours on end. Still, he recognized important work when he saw it, and in Riemann he saw some of the greatest promise he had ever witnessed. The two were kindred souls.

Riemann, the son of a poor Lutheran pastor, was originally planning to study theology, but his interest and abilities in mathematics helped turn him away from that path. At the age of twenty, he attended the University of Göttingen and sat in on some of Gauss's lectures. After a few years in Berlin, he returned to Göttingen, where he would spend the rest of his career.

In the early 1850s, Riemann was working on what is known as a habilitation—the highest academic qualification someone in Europe in the nineteenth century could receive. Habilitations were always taken after a doctorate was completed, which required Riemann to present a second dissertation.

By the rules of the university, it fell to Gauss to approve the sub-ject of the second dissertation. Riemann had to submit three topics: a first choice and two alternatives. Usually the way that this worked was that the adviser (Gauss) would simply give the green light to the chosen subject of the person submitting the dissertation. By these standards, Gauss could have, and probably should have, selected the subject that Riemann had selected as his first choice. That was the custom. But in this case, Gauss made the unusual choice of selecting one of Riemann's alternative dissertations—on the foun-dations of geometry.

The culmination of this choice came when Riemann gave a lec-ture to his fellow faculty members at Göttingen on June 10, 1854, shortly before Gauss died. Riemann's thesis was titled "On the Hypotheses Which Lie at the Basis of Geometry." Gauss attended the lecture, which set in motion a change in mathematics that con-tinues to unfold today. Through this and the work that Riemann did over the next decade, he created a whole new form of non-Euclidean geometry that was complementary to that of Lobachevsky and Bolyai. He did this by once again taking aim at the fifth postu-late and asking what would happen if it were not true.

In Lobachevsky's non-Euclidean geometry, there were infinite lines that could be drawn parallel to a given line through a point not on that line. In Riemann's non-Euclidean geometry, there was no such thing as a parallel line. And because his lines were never parallel, there was no need for the fifth postulate anymore. But Riemann did not stop there. He also took aim at Euclidean geome-try by rejecting another of the first five postulates. He set his sights on the second postulate, which states, "Any straight line segment can be extended indefinitely in a straight line."

Riemann's non-Euclidean geometry, or "elliptical geometry" as it became known, rejects parallel lines and the idea that any straight line can go on indefinitely. Lines in his new geometry were never infinite but always of a finite length. This was a profoundly new development because it drew into question the fundamental idea that space itself is infinite.

The easiest way to imagine Riemann's new type of non-Euclidean geometry is to picture a sphere. The surface of a sphere is one of constant positive curvature, and lines drawn on it are not straight but curved. You can easily convince yourself that lines are always finite in length by drawing a few of them. The geometry of the sphere dictates that any line that follows the curvature of the sphere will connect back up with itself and make a great circle— essentially cutting the sphere in half.

Now, as far as parallel lines go, it is relatively easy to convince yourself that they do not exist. There is no way to draw two lines encircling a sphere that do not cross. Any two great circles that you can draw on the sphere will overlap—cross—and since any two lines always cross, parallel lines do not exist.

If that is not easy to visualize, consider an everyday sphere, like an orange. You can cut an orange exactly in half in any number of ways, and when you do you always wind up with two equal halves. But how can you make two cuts through the center of an orange without having your cuts overlap? You can't. Regardless of how you place your second cut, you will always wind up with four pieces.

More than just inventing a third type of geometry, what Riemann did was to reach to the very roots of the discipline and create a general foundation upon which both Euclid's geometry and both types of non-Euclidean geometry could be built. He showed that Euclidean, Lobachevskian, and his own geometries were but specific examples of a more general system. In doing so, he created a rigorous mathematical formulation that allows it to be used more easily.

The nature of the space, according to Riemann, differs depending on its curvature. For him, space could have infinitely many structures. Euclidean geometry is merely a particular example of this. "Space is only a special case of the three dimensions," he wrote. What we would think of as normal space, based on Euclidean geometry, is simply a special case of geometry where there is zero curvature. The "hyperbolic" non-Euclidean geometry created by János Bolyai and Lobachevsky is a special case where the space has a constant negative curvature. Riemann's own "elliptical" geometry is a special case where the space has a constant positive curvature.

Riemann generalized the work Gauss had done years earlier and extended his studies of the intrinsic properties of a surface to any surface in any type of space. Moreover, he separated the concepts of shape and space by generalizing geometry to apply to any space of a given number of dimensions. In this generalized geometry, shapes are defined in absolute terms and not in terms of an arbitrary, external three-dimensional space. He did this by defining n-dimensional space in which any number of coordinates like x, y, and z can be used. Those n coordinates define the position of a point. Adding to the concept of a general n-dimensional space, Riemann showed how individual elements of distance can be represented on it and how measurements can be made.

In his lecture, Riemann outlined the position that three-dimensional representations of space are arbitrary. "An assumption," he stated, "which is developed by every conception of the outer world." By creating n-space, the three types of geometries grew to infinity types under Riemann. Geometry was no longer simply the study of shapes in space but the study of transformations and symmetry. Riemann's extended geometry had a rigorous mathematical formulation that allowed it to be used more easily.

Riemann also introduced the possibility of the finiteness of space. He realized that the curvature of space in the universe would have profound implications for its scale. Is the universe finite in extent, or does it go on into infinity? "If we assume independence of bodies from position and therefore ascribe to space constant curvature, it must necessarily be finite," he wrote.

The most fundamental conceptual breakthrough Riemann made was to determine that bodies in a physical space are not simply occupants of this space but actors that bend and shape the space itself by their very presence. In doing so he anticipated the central concept and laid out the mathematical foundation of Einstein's general relativity theory by more than sixty years and even suggested that space could be measured by its physical masses. This insight was astounding considering that Riemann was a classical physicist working more than fifty years before the advent of relativity.

Riemann was an exceptional genius whose work was incredibly complicated. When he died from tuberculosis in 1866, his work was still relatively undiscovered. By the end of the nineteenth century there still was no popular treatment of it. But there was a slow growth of mathematicians who appreciated the revolution he had helped bring about—as well as the work of Gauss, Lobachevsky, and János Bolyai—that was beginning to come into being.

A letter that Gauss sent to a friend praising his Russian contemporary surfaced and was printed in Germany. This letter introduced many in European mathematical circles to the name Lobachevsky for the first time. Reading that Gauss appreciated his work motivated other mathematicians to seek out the work themselves. It was only a matter of time before people rediscovered the earliest works on non-Euclidean geometry and began to appreciate them.

In 1867, a professor in Dresden named Richard Baltzer discovered and advanced the works of János Bolyai and Lobachevsky in a textbook called *Elements of Mathematics*. In this book, Baltzer incorporated references to their works and to the new ideas of non-Euclidean geometry. Guillaume-Jules Hoüel read Baltzer's work and pushed hard to publish it and several other works of non-Euclidean geometry in French. He translated Lobachevsky into French in 1866, along with some of Gauss's correspondence, and published this in France. He wrote in the preface, "Mr. Richard Baltzer, in the second edition of his *Elements of Geometry*, at the beginning, introduces these exact notions in the place that they must occupy." He translated János Bolyai's work a few years later and wrote the first biography of him, also in French.

The 1860s and 1870s were an exciting time for non-Euclidean geometry. From just a few people who appreciated the subject, suddenly there were more than a handful of mathematicians who were interested and able to understand it as well. Besides Hoüel in France, an English mathematician named William Kingdon Clifford was thinking about and working out what non-Euclidean geometry could mean for matter and energy in space. He published an important work in 1870 called *On the Space Theory of Matter*. In Belgium the military man and mathematician Joseph Tilly was also studying the subject, as was Hermann von Helmholtz in Germany. They found in Lobachevsky a body of work that had been developed over decades and was ready to be examined and extended.

Helmholtz was a strange product of his times. Interested in mathematics but unable to afford a university education, he instead went to medical school on a government scholarship. For this he had to serve for nearly a decade as an army doctor. But he is much more famous today for his impact on mathematics. In 1868, he wrote a paper that embraced Riemann's work and helped cement his position in the mathematical world as an acceptable way of looking at geometry.

In 1867, an Italian mathematician named Eugenio Beltrami wrote treatments of János Bolyai and Lobachevsky that appeared in Italian publications. Within a few years, these were translated into

French. Beltrami was only one of several European mathematicians who were translating and commenting on the works of János Bolyai, Lobachevsky, and Riemann, but he played an especially important role because he succeeded in popularizing non-Euclidean geometry. Beltrami's work made non-Euclidean geometry more accessible and also more believable.

Beltrami was a troubled student who was noted for acting out. He had a rough run as a scholar early in his career, being dismissed from one of his first jobs after only two months for "political" reasons. But just a few years later he was named extraordinary professor of the University of Bologna, and from there he rose to the highest station that a scientist in Italy could achieve. It was in this position that he began to work on geometry. He had known Riemann, and the two had met frequently. Like Riemann, Beltrami was deeply influenced by Gauss.

In his work, Beltrami proved that Euclidean and non-Euclidean geometry are logically consistent. This was only the beginning. What Beltrami did next was to create a surface that was a convenient and easily visualized way of representing non-Euclidean geometry in ordinary three-dimensional space.

Beltrami's invention was a geometrical construct known as a pseudosphere. In a pseudosphere, the straightest lines run out to infinity. Also, numerous lines can be drawn parallel to a given line. Beltrami's pseudosphere was a significant breakthrough because it helped change the perception of non-Euclidean geometry from something that was seen as completely abstract and imaginary to something that was abstract but could be represented in real-world terms. He made the subject "intuitively true" by conceiving of these surfaces.

Richard Dedekind, a former student of Gauss's who had finished his habilitation degree at the same time as Riemann, published Riemann's 1854 lecture in 1868. Beltrami read this with interest. He translated the lecture and commented on it. By 1870, Beltrami was publishing work on non-Euclidean geometry that combined the ideas of Riemann with Gauss, Lobachevsky, and János Bolyai.

In Germany, Felix Klein did further work on the subject, and by 1871 he and Beltrami had succeeded in making non-Euclidean geometry a mathematically respectable subject. Klein fully developed the idea of a general geometry as being about the invariant properties of a group of defined transformations. His work fundamentally advanced non-Euclidean geometry from a rigorous, if fanciful, subject into something that was on equal footing with Euclidean geometry. Klein also gave Gauss so much credit for non-Euclidean geometry that it would become his standard due. "By the weight of his authority," said Klein, "he first brought this intellectual creation, heavily contested at first, to general attention and ultimately to victory."

Klein coined the phrase *hyperbolic geometry* to describe what János Bolyai and Lobachevsky had invented. He drew upon the Greek word *hyperballein*, which means "to exceed." Likewise, he called the geometry that Riemann invented *elliptic* from the Greek *elleipein*, or "to fall short."

After all these wheels were in motion, a succession of mathematicians from the next generation embraced and furthered work in the field. One of the most noteworthy was Henri Poincaré. He was important because he helped to settle the philosophical crisis that arose from the now recognized existence of multiple geometries.

Poincaré reconciled Euclidean and non-Euclidean geometries by reasoning that they were not incompatible competing systems. For Poincaré, the question of which of these multiple geometries was true or false was the wrong question. The real question was which geometry was more convenient than another. He once wrote, "The axioms of geometry are neither synthetic judgments *a priori* nor experimental facts. They are conventions."

Poincaré lent an air of credibility to non-Euclidean geometry. He was from a highly distinguished French family. His father was a professor of medicine, and his cousin served as president of the French Republic during World War I. He was uniquely suited to become the ambassador of non-Euclidean geometry, which was exactly what he became when he took on the task of writing essays on the subject. Poincaré was practically a celebrity scientist when he set about popularizing non-Euclidean geometry, and it was thanks to him that the subject became universally interesting. He wrote essays at the turn of the century that were widely influential in making non-Euclidean geometry accessible far and wide.

Thanks in large part to Poincaré's efforts, by the dawn of the twentieth century, a critical mass of artists, writers, mathematicians, and scientists began to wrestle with the idea of non-Euclidean geometry. It would breathe new life into mathematics and give birth to everything from M. C. Escher's engravings to Lewis Carroll's literature. H. P. Lovecraft turned non-Euclidean geometry into the subject of horror and insanity in some of his stories. Non-Euclidean geometry allowed artists to begin to depict forms not as they are in reality—in nature—but as they are in a consistent representational space. Just as the Renaissance witnessed the awakening of a renewed appreciation of mathematics and geometry, the late nineteenth and early twentieth centuries saw a new interest in art forms based on higher dimensional geometries. Modern art was born when appreciation of non-Euclidean geometry was at its peak, and the influence of the one on the other may have been felt in the work of Marcel Duchamp, who read Poincaré's essays and was deeply influenced by them.

What is not so clear is the effect that non-Euclidean geometry had on the modern painters Pablo Picasso and Georges Braque, the founders of cubism. Some claim that non-Euclidean geometry and relativity inspired the modern artists to invent cubism as the artistic realization of the new subject. Perhaps they can trace the explorations of the fourth dimension by artists like Salvador Dalí to non-Euclidean geometry and relativity. Others are not so convinced. One artist who was influenced by non-Euclidean geometry but was not himself convinced of its validity was Lewis Carroll, who was inspired by the curvature of non-Euclidean geometry into an absurd world in his novel *Alice's Adventures in Wonderland*. "Lewis Carroll" was really the alter ego of the Reverend C. L. Dodgson, a mathematician whose work encompassed obscure tricks like finding a method for determining the day of the week of any date in a given year or a trick for dividing a given number by 11 or 9.

Dodgson was a champion of elementary geometry and of Euclid, and he was strongly against non-Euclidean geometry. He attacked the supporters of non-Euclidean geometry in his book *Euclid and His Modern Rivals* in 1879. In this he attempted to show how the two-thousand-year-old subject was logically superior to any other.

He was not alone. Many mathematicians at the end of the nineteenth century made a case against non-Euclidean geometry by propping Euclid and his two-millennium-long tradition up against the upstart non-Euclideans. Many of them thought of non-Euclidean geometry as a logical exercise—something like the creative mental equivalent of a science-fiction novel—a fiction without any true application.

People interested in non-Euclidean geometry could toy with their imaginary geometries all they wanted, but in the end, what did it really matter? It was the first contradiction-free mathematical system with no connection to reality. Was it something one could actually use? Was it interesting and intellectually invigorating? Maybe, if you're into that sort of thing.

Euclidean geometry was the geometry of our everyday, real experience. In Euclidean geometry, straight lines go off to infinity. Parallel lines are those that lie in the same plane but do not meet. We experience parallel lines traveling off into the distance, like the two white lines in the middle of a road headed for the vanishing point. We experience ordinary triangles whose angles add up to 180 degrees. We experience lines that are not curved and rectangles and all those things that describe reality and that non-Euclidean geometry maddeningly challenges.

The good news of non-Euclidean geometry is that a point is still a point and a line is still a line. In the non-Euclidean geometry invented by Lobachevsky and János Bolyai, the three angles of a triangle do not add up to 180 degrees, and there is no such thing as similar triangles. The Pythagorean theorem is no longer true. There are infinitely many parallel lines. The strangest result of non-Euclidean geometry is that if you draw a straight line and then draw another line that is everywhere equidistant from the first line, that second line will be curved.

All these things are consequences of solving the mystery of the fifth postulate—as Lobachevsky and János Bolyai did—by declaring it to be false. The fifth postulate is the difference between normal geometry and non-Euclidean geometry, and throwing it away opened a mathematical Pandora's box. Clearly not everyone would be willing to accept this development!

The greatest ally Euclid had was our corner of the universe and the fact that we live in it. The problem with living on Earth is that we are biased into thinking that space in the universe is as we take it to be from our experience. Euclid's geometry seems like a right and natural representation of space because it exactly resembles our experience. Geometry is what we see when we look around us. Euclidean geometry works—exactly the reason that the Egyptians invented it and also the reason that it took so many years to solve the fifth postulate. Nobody ever doubted it was true because it made so much sense.

The greatest enemy non-Euclidean geometry ever had was Immanuel Kant. Though he died years before non-Euclidean geometry was invented, there was no greater authority on the philosophical argument against the subject. Kant advocated the necessity of space and strongly endorsed Euclid in his *Critique of Pure Reason.* Kant sought to make space a subjective phenomenon that was necessary for experience. But space was to Kant the space of our everyday experience—defined by ordinary, Euclidean geometry. By making Euclid's geometry philosophically necessary, Kant gave it a great boost among mathematicians who sought to argue against the validity of non-Euclidean geometry. Thanks to him, many were reluctant to accept it

By the beginning of the twentieth century, these objections began to melt away, and non-Euclidean geometry came to be more accepted—even if it was not the geometry of our experience. Ironically, some of Kant's defenders even claimed that his philosophy anticipated non-Euclidean geometry. Nevertheless, there were still staunch opponents at the turn of the century. Some of the top science and math journals chronicled the pitched battle for the soul of geometry. As the lead editorial in an 1895 issue of the *American Mathematical Monthly* said, there was no shortage of opposing opinions. "Every great advance in science, every great discovery in nature, and every great invention has had its crowd of ridiculers; non-Euclidean geometry is no exception," it read.

One of the main critics of non-Euclidean geometry at the turn of the century was an American mathematician named John Lyle. He wrote many attacks of the new subject, often blistering and cynical. These attacks focused on the new geometry's strange nature. "I would like to see a Lobachevsky triangle before I die," he penned smugly in 1894, "but if it is impossible to materialize the nondescript in 'space as we know it' I suppose that I will have to forego the pleasure."

Lyle and the other opponents of non-Euclidean geometry were perfectly willing to admit that the mathematics and the subject were interesting, but they were thoroughly convinced it was so abstract as to have no connection to reality—and because of that no validity.

A strangely aggressive and offensively snide professor named Eugen Dühring made the most blistering attacks on Gauss and non-Euclidean geometry. He denounced the new geometry, dismissed Gauss as a narrow-minded fool, and decried what he saw as a culture of Gauss worship. He called non-Euclidean geometry the abortive product of a deranged mind (Gauss's). "His megalomania rendered it impossible for him to take exception to any tricks that the deficient parts of his own brain played on him, particularly in the realm of geometry," Dühring once wrote. "Thus he arrived at the pretentiously mystical denial of Euclid's axioms and theorems, and proceeded to set up the foundations of an apocalyptic geometry not only of nonsense but of absolute stupidity."

In the end, non-Euclidean geometry won. It was fostered by the mathematicians who followed after János Bolyai, Gauss, and Lobachevsky. A new appreciation for their works bloomed. By the end of the century, even some textbook writers began to be influenced by the ideas of non-Euclidean geometry and rewrite their books to reflect the new way of thinking about space in general.

One measure of this victory is that by the end of the century, the popular art of attempts to prove the fifth postulate faded away completely. By then, no respectable mathematician would try to prove it. Some even decried the idea of trying as preposterous—positively hopeless. Saying something like that anytime in the previous two thousand years would have been considered mathematical heresy.

The development of non-Euclidean geometry was intellectually one of the bravest acts in history because it meant going against more than two millennia of established doctrine by rejecting the fifth postulate, a logical statement that represents space as we perceive it in favor of an otherworldly universe where lines are curved.

But if non-Euclidean geometry was the solution to the greatest mathematical mystery of the ages, what difference did it make? By the end of the nineteenth century, geometry was no longer taught as it once was—straight from the ancient text of Euclid's *Elements*— but it was still more or less Euclidean. And the shape of the world as people saw it was still Euclidean: three-dimensional, rectilinear, and straight. But the importance of non-Euclidean geometry should not be measured by the fact that Euclidean geometry persisted but by its importance in the world of the twentieth century—as one of the important foundations of modern mathematics and cosmology.

Non-Euclidean geometry was a major advance in abstract thinking. Fundamentally it separated mathematics from physics. No longer was geometry a necessary representation of reality but a convenient one. And no longer was geometry held hostage by reality. Mathematicians were now free to pursue a logical mathematical system that was divorced from reality. More important, physicists were no longer mathematically bound to the world of our observation like Prometheus on the rocks.

By the turn of the century, textbooks were being printed describing the methods of the new geometry. Mathematicians everywhere were reconsidering not only the fifth postulate but geometry as a whole, examining it for any logical inconsistencies or assumptions. The ultimate expression of this rethinking was an 1899 book by David Hilbert called *Foundations of Geometry*, which completely reinvented Euclid's *Elements* in a more rigorous way.

There were still some mathematicians who objected, but the field in general slowly began to be populated by mathematicians who saw their art as a way of representing reality, not reality itself. The most important aspect of non-Euclidean geometry was not whether it was correct, useful, or even easy to understand. The most important thing was that it brought an expansion of possibilities that may still be untapped today. It was a critical milestone because it helped set the stage for the development of modern physical theories. No longer was mathematics to be considered a natural science by necessity. Mathematics is an abstraction of reality.

Today geometry is a subject that makes hypotheses and asserts consequences that follow from them. It is not bound by the requirement that the mathematical objects it deals with are necessary. The discovery of non-Euclidean geometry gave mathematicians a sense of the relative rather than the absolute. The truths of geometry and mathematics in general need not be absolute. Relativity is permissible. This fact was discovered shortly after the turn of the century by Albert Einstein, who read the works of Hermann von Helmholtz together with his girlfriend Mileva Maric in 1899 before they were married.

In the end, the story of non-Euclidean geometry parallels one of the greatest changes in the history of mathematics. There was a time when the best mathematicians of an age could understand and contribute to all the fields of the discipline that existed. Over the nineteenth century, as non-Euclidean geometry was being invented, mathematics underwent a transition. At the beginning of the century there were many men (and at least one woman) who understood the entire field. By the end of the century, only one person (Henri Poincaré) was so well versed—and nobody after him would ever be again.

The story of non-Euclidean geometry is also one of strange parallels. It all started with trying to prove the fifth postulate, the idea that lines that are parallel will never meet. For thousands of years, mathematicians followed the same path by trying to prove the idea, fail, or give up trying. Just like parallel lines, the lives of Gauss, Lobachevsky, and János Bolyai ran alongside one another.

Each of them made his own journey of discovery, finding the unexpected truth of geometry and the fifth postulate. They all discovered that the secret to solving the mystery of the fifth postulate was not to prove it but rather to reject it outright. Gauss, Lobachevsky, and János Bolyai all discovered that they didn't need to find a way around the fifth postulate, and they found this independent of one another. They knew nothing of the work of the others when they made their own journeys, which is hardly surprising. They were of the same age but they were not of the same generation. Gauss was in his twenties and Lobachevsky was of elementary school age when János Bolyai was born.

They were different men of different ages in distant countries. Theirs were parallel lives, and like the parallel lines with which they concerned themselves, the three never met.

17

The Curvature of Space

The space and time of physics are merely mental scaffolding in which for our own convenience we locate the observable phenomena of nature.
—Sir Arthur Eddington

When the English physicist Sir Arthur Eddington set about organizing an expedition to view the full eclipse of the sun due to take place in May 1919, the first question was where to go. No place in Great Britain would do—nowhere in Europe in fact. To get a good glimpse of the full eclipse, one had to stand directly under the path the moon's shadow swept out as it passed between Earth and the sun.

Timing and location were everything, because the moon would sweep in front of the sun somewhere over central Africa, shade a crescent path across the Atlantic Ocean, and plunge a swath of northern Brazil into semidarkness before finally moving out of the way of the sun a few minutes later. To view the full eclipse, Eddington and his expedition would have to get right along this crescent path. He chose to go to a small island off the coast of West Africa, and he caught a good view of the eclipse. He even captured a stunning image of the sun's corona on film.

He was not interested, however, in the sun or the moon or the spectacle of seeing the sun's corona that day. What he was doing was testing the location of a star on the other side of the sun. Light from this star would normally be impossible to see because the light of the sun is much too intense. But with most of the direct light of the sun blotted out during the total eclipse, the star would be visible. The question was, *where?*

According to the classical definition of space, the universe is Euclidean and three-dimensional, and light from this distant star would be expected to travel along a straight line. In other words, the star should appear exactly where it would be expected to appear. But in 1919 there was another theory about the nature of space—Einstein's theory.

Albert Einstein had postulated the idea of curved space-time in his general theory of relativity. He was studying the geometry of space and the mechanical force of gravitation and found them to be non-Euclidean in relativity. According to general relativity, which he established by 1915, light travels along a non-Euclidean line—the shortest path between two points was not a Euclidean line but rather a curved non-Euclidean one.

The crucial element that allowed Einstein to apply non-Euclidean geometry to space like this was the full equation of electrodynamics, which had been worked out by James Maxwell in the second half of the nineteenth century. Maxwell showed that an electromagnetic wave travels through space at a speed that is the ratio of two constants from electricity and magnetism. This meant that the speed of light was itself a constant. The constancy of the speed of light brought up a complication: regardless of how fast you are moving away from or toward a light source, the speed of light remains unaltered. How could the speed of light be constant regardless of motion?

The physicist Hendrik Lorentz addressed this issue by suggesting that space can contract. He determined that the length of a moving body would contract in the direction of motion. Lorentz and Einstein independently worked out the equations for these transformations, and Einstein applied them to special relativity. The term "relativity" refers to the invariance in certain physical properties in different frames of reference—like the speed of light. Under relativity, other properties must undergo transformations so as to leave the other, invariant physical properties unaltered.

After working out special relativity in 1905, Einstein took about a decade to develop his general theory of relativity, which brings in the effect of gravity. The premise of general relativity is that geometry is a physical entity that can act on matter and be acted upon by matter. The geometry of the universe, according to general relativity, is determined by the matter it contains. The strange outcome of general relativity is that gravity curves space-time. Or rather, gravity, according to Einstein's theory, is the result of curved space-time, which could be represented by non-Euclidean geometry.

Space itself may be Euclidean if nothing is there, but in the presence of mass it is warped or curved. Near a massive object that has an enormous gravitational pull, the geometry of space becomes curved. As the physicist John Archibald Wheeler put it, "Space tells matter how to move and matter tells space how to curve."

In developing general relativity, Einstein embraced non-Euclidean geometry as the mathematical tool to describe this curvature. Specifically, he turned to the non-Euclidean geometry developed by Bernhard Riemann and influenced by Carl Gauss's work as the mathematical system that would form the basis for his new theory. The general theory of relativity is an impressive one. Under certain conditions space-time is flat and three-dimensional. At other times it is curved. Deviations from Euclidean geometry appear as a gravitational field.

In relativity, the properties of real space and time correspond to Riemann's geometry rather than Euclid's, and this is what the 1919 expedition proved. Eddington's expedition demonstrated Einstein's formulations to be correct. When the photographs taken during the total eclipse were compared to photographs taken when the sun was not in front of the same stars, their positions were different. The light from that distant star, visible when the sun was fully eclipsed, was in an apparent location other than where it was expected to be. The reason for this deviation was that the path followed by the light emitted from the star was bent. Space-time was curved. The observations showed that mass affected space-time.

Einstein became famous overnight. Some claim that this effect was enhanced by the fact that humanity as a whole had just endured one of the most trying times imaginable. World War I had visited horrors previously unimaginable. Directly after its end, the Spanish flu had brought equally unimaginable suffering on the home front. But this is not to take anything away from what Einstein did. His was a discovery worthy of celebration in any era.

Gauss probably never knew how far non-Euclidean geometry would take physics. It continued to have applications in cosmology and celestial mechanics. Another important development in the early twentieth century was Edwin Hubble's observations of the expanding universe, a tantalizing demonstration of the big bang theory and one that is consistent with general relativity and its non-Euclidean geometry formulations. The big bang model is based on mathematical solutions of the equations of general relativity.

Einstein, who as a student studied Gauss's theory of surfaces, called Gauss's importance overwhelming. "If he had not created his geometry of surfaces, which served Riemann as a basis, it is scarcely conceivable that anyone else would have discovered it," he said. "The importance of C. F. Gauss for the development of modern physical theory and especially for the mathematical fundament of the theory of relativity is overwhelming indeed."

We think of Gauss today as a mathematician—one of the greatest of all time. If he is remembered as more than that, it's because of his mathematical skills and influence. These were so great that he fully transcended the mathematics of his day and saw things of incalculable complexity. He had a curious and probing mind, and he tinkered with numbers the way that a great writer might play around with words or a great composer with musical notes. For Gauss, numbers and equations were his instruments and notes of creative expression. And the quality of his expression was almost unequaled in the history of mathematics.

Understanding an overwhelming genius like Gauss's is a difficult thing. And this is made all the more difficult by the fact that he lived a long, full, and seemingly normal life free from the tragedies and tributations of people like Archimedes or Galileo; free from the strange interests and controversies surrounding Isaac Newton; and free as well from the persecution that some of his contemporaries felt, living as they were in politically turbulent times in Europe.

If we are meant to understand his genius at all, perhaps the best approach is not another microscopic tissue probe or another gross reexamination of Gauss's pickled brain sitting on a shelf somewhere in Germany. Perhaps the only way to approach this understanding is by looking at the mathematical ideas that his mind produced—his great works—both those that he published, his books that have stood the test of time, and those ideas that he never told anyone about, such as non-Euclidean geometry.

Notes

1 A Mathematician's Waterloo

The chapter epigraph is taken from G. Waldo Dunnington, *Gauss: Titan of Science* (Washington, DC: Mathematical Association of America, 2004), 120.

3   *The French capital attracted*   Ernest W. Brown, "The History of Mathematics," *Scientific Monthly* 12 (May 1921): 400; and Jesse A. Fernandez Martinez, "Sophie Germain," *Scientific Monthly* 63 (October 1946): 257.

3   *Lagrange was the elder*   Brown, "The History of Mathematics," 385–413; E. T. Bell, *Men of Mathematics* (New York: Simon and Schuster, 1986); and George P. Meade, "Youthful Achievements of Great Scientists," *Scientific Monthly* 21 (November 1925): 522–532.

4   *By the age of twenty-five*   Meade, "Youthful Achievements of Great Scientists," 524.

4   the French Revolution   Michael Rapport, *Nineteenth-Century Europe* (New York: Palgrave Macmillan, 2005), 12.

5   *an Italian by birth*   Raymond C. Archibald, "History of Mathematics after the Sixteenth Century," *Mathematics*, January 1949, 48.

5   *Frederick the Great*   G. Waldo Dunnington, "The Historical Significance of Carl Friedrich Gauss in Mathematics and Some Aspects of His Work," *Mathematics News Letter* 8 (May 1934): 175; and Bell, *Men of Mathematics,* 165.

5   *When he was facing the guillotine*   George Bruce Halsted, "The Message of Non-Euclidean Geometry," *Science* 19 (March 11, 1904): 401.

5   *Napoleon . . . a keen interest in Lagrange*   Bell, *Men of Mathematics,* 153, 170.

6   *Such was the mood*   Florence Lewis, "History of the Parallel Postulate," *American Mathematical Monthly* 27 (January 1920): 18; also Jeremy Gray, *Janos Bolyai, Non-Euclidean Geometry, and the Nature of Space* (Cambridge: MIT Press, 2004), 40. There are conflicting accounts on this. One held that the speech was given in 1806.

6   *The oldest conundrum in mathematics*   George Bruce Halsted, "Bibliography of Hyper-Space and Non-Euclidean Geometry," *American Journal of Mathematics* 1 (1878): 261–276.

8   *"Il faut que j'y songe encore"*   Lewis, "History of the Parallel Postulate," 18; also Gray, *Janos Bolyai, Non-Euclidean Geometry, and the Nature of Space,* 40.

8   *Euclid's* Elements *was irreproachable*   V. Kagan, *N. Lobachevsky: His Contribution to Science* (Moscow: Foreign Language Publishing House, 1957), 14.

9   *one of the great achievements of the human mind*   Shiing-Shen Chern, "What Is Geometry?" *American Mathematical Monthly* 97 (October 1990): 679.

10   *so obsessed with numbers*   Dunnington, *Gauss: Titan of Science*, 219; also W. K. Bühler, *Gauss: A Biographical Study* (Berlin: Springer-Verlag, 1981), 81; and G. Waldo Dunnington, "The Sesquicentennial of the Birth of Gauss," *Scientific Monthly* 24 (May 1927): 411.

10   *do cube roots in his head*   Meade, "Youthful Achievements of Great Scientists," 524.

11   *he enjoyed using mathematical tables that were cheap*   Dunnington, *Gauss: Titan of Science*, 258.

11   *stretched his work across every field of mathematics*   Ibid., p. 252.

11   complex numbers   Dunnington, *Gauss: Titan of Science*, 40, 112.

11   *"Your* Disquisitiones"   Ibid., p. 208.

11   *"Thou, Nature, art my goddess"*   Ibid., p. 207.

12   Johann Friedrich Carl Gauss   M. B. W. Tent, *Carl Friedrich Gauss: Prince of Mathematics* (Wellesley, MA: AK Peters, 2006), 2.

12   *Gebhard Dietrich Gauss*   Dunnington, *Gauss: Titan of Science*, 11; also Bühler, *Gauss: A Biographical Study*, 6; G. E. Burch, "Carl Friedrich Gauss—A Genius Who Apparently Died of Arteriosclerotic Heart Disease and Congestive Heart Failure," *A.M.A. Archives of Internal Medicine* 101 (April 1958): 825.

12   *Gauss's mother*   Dunnington, "The Sesquicentennial of the Birth of Gauss," 403.

12   *"half citizens"*   W. K. Büler, *Gauss: A Biographical Study* (New York: Springer-Verlag, 1981), 5.

13   *his family was so poor*   Dunnington, *Gauss: Titan of Science*, 231.

13   *He didn't know when his birthday was*   Ibid., p. 69.

13   *Gauss supposedly interjected*   Burch, "Carl Friedrich Gauss—A Genius Who Apparently Died of Arteriosclerotic Heart Disease and Congestive Heart Failure," 825.

14   St. Katherine's school   András Prékopa and Emil Molnár, *Non-Euclidean Geometries* (New York: Springer, 2006), 6; also Rapport, *Nineteenth-Century Europe*, 8, discusses schools in Germany during the eighteenth century.

16   Mr. Büttner   Burch, "Carl Friedrich Gauss—A Genius Who Apparently Died of Arteriosclerotic Heart Disease and Congestive Heart Failure," 826.

16   Johann Christian Martin Bartels   Bühler, *Gauss: A Biographical Study*, 7.

17   Eberhard August Wilhelm von Zimmerman   Dunnington, *Gauss: Titan of Science*, 11.

18   *to foster genius within their realm*   D. J. Struik, "Outline of a History of Differential Geometry II," *Isis* 20 (November 1933): 161. Perhaps the most famous of these was Germany's great mathematician Gottfried Leibniz, who died sixty years before Gauss was born. Leibniz had been a court mathematician and much more, serving the dukes of Hanover for decades and helping them immensely.

18   Zacharias Dase   Archibald, "History of Mathematics After the Sixteenth Century," 52.

19   *"the humble pursuits of trade for those of science"*   *"Obituary Notices of Deceased Fellows,"* *Proceedings of the Royal Society of London* 7 (1854–55): 590.

20   the University of Göttingen   Bühler, *Gauss: A Biographical Study*, 15; also Dunnington, "The Historical Significance of Carl Friedrich Gauss in Mathematics and Some Aspects of His Work," 176; and Dunnington, *Gauss: Titan of Science*, 23.

20   Benjamin Franklin   Dunnington, *Gauss: Titan of Science*, 280.

20   *a seventeen-sided figure*   Archibald, "Gauss and the Regular Polygon of Seventeen Sides," *American Mathematical Monthly* 27 (July–September 1920): 324; also Archibald, "History of Mathematics After the Sixteenth Century," 51; and Bühler, *Gauss: A Biographical Study,* 28.

22   *the most successful textbook of all time*   Richard Trudeau, *The Non-Euclidean Revolution* (Boston: Birkhäuser, 1987); also Walter C. Eells, "Discussions: The Ten Most Important Mathematical Books in the World," *American Mathematical Monthly* 30 (September–October 1923): 319; and P. H. Daus, "The Founding of Non-Euclidean Geometry," *Mathematics News Letter* 7 (April–May 1933): 12.

22   *analytical thinking and the scientific method*   D. Fraser Harris, "The Influence of Greece on Science and Medicine," *Scientific Monthly* 3 (July 1916): 52.

23   *native son of Tyre*   J. J. O'Connor and E. F. Robertson, "Euclid of Alexandria," from the Web site of the School of Mathematics and Statistics at the University of St. Andrews, Scotland, www-history.mcs.st-andrews.ac.uk/Biographies/Euclid.html, last accessed May 2007.

23   *Plato was not interested in experimentation*   William Salant, "Science and Society in Ancient Rome," *Scientific Monthly* 47 (December 1938): 530; also George Sarton, *Introduction to the History of Science,* vol. 1: *From Homer to Omar Khayyam* (Baltimore: Williams and Wilkins, 1927), 92.

23   *Aristotle was even more influential*   Roberto Torretti, *Philosophy of Geometry from Riemann to Poincaré* (Dordrecht, Holland: D. Reidel Publishing Co., 1984), 5.

2 The Strange Vegetarian Cult and Mathematics

The chapter epigraph is taken from Kenneth Sylvan Guthrie, *The Pythagorean Sourcebook and Library* (Grand Rapids, MI: Phanes Press): 73.

25   *son of Apollo*   Charles Kahn, *Pythagoras and the Pythagoreans: A Brief History* (Indianapolis: Hackett), 5.

25   *Pythagoras of legend*   Kenneth Sylvan Guthrie, *The Pythagorean Sourcebook and Library* (Grand Rapids, MI: Phanes Press, 1988); also Kahn, *Pythagoras and the Pythagoreans.*

25   *The fishermen were amazed*   Ibid, p.65.

26   *of the man we know little*   C. R. Eastman, "Notes on the History of Natural Science," *Science* 21 (March 31, 1905): 517.

26   *Pythagoras was one of the most*   Guthrie, *The Pythagorean Sourcebook and Library,* 63.

26   *"bristled" with geometry*   Roger Osborne, "Some Historic and Philosophic Aspects of Geometry," *Mathematics* 24, no. 2. (November–December 1950): 77.

27   Egyptian calendar   J. J. O'Connor and E. F. Robertson, "An Overview of Egyptian Mathematics," from the Web site of the School of Mathematics and Statistics at the University of St. Andrews, Scotland, www-groups.dcs.st-and.ac.uk/~history/HistTopics/Egyptian_mathematics.html, last accessed July 2007.

27   *All this is revealed in an ancient papyrus*   G. A. Miller, "Some Fundamental Discoveries in Mathematics," *Science* 17 (March 27, 1903): 496–499; also J. J. O'Connor and E. F. Robertson, "Mathematics in Egyptian Papyri," from the Web site of the School of Mathematics and Statistics at the University of St. Andrews, Scotland, www-groups.dcs.st-and.ac.uk/~history/HistTopics/Egyptian_papyri.html, last accessed July 2007.

28   *The Babylonians knew about certain geometrical theorems*   O'Connor and Robertson, "An Overview of Babylonian Mathematics"; also Raymond J. Seeger, "The

Exact Sciences in Antiquity by O. Neugebauer," *Science* 117 (March 6, 1953): 257; H. E. Buchanan, "The Development of Elementary Geometry," *Mathematics News Letter* 3 (January 1929): 9; and Roberto Torretti, *Philosophy of Geometry from Riemann to Poincaré* (Dordrecht, Holland: D. Reidel Publishing Co., 1984), 2.

28   *Some have even questioned whether the Egyptians*   Seeger, "The Exact Sciences in Antiquity by O. Neugebauer," 257; also Klaus Galda, "An Informal History of Formal Proofs: From Vigor to Rigor?" *Two-Year College Mathematics Journal* 12 (March 1981): 126.

29   *Another Egyptian treatise*   Buchanan, "The Development of Elementary Geometry," 10.

29   *an art as much as a science*   "Striving for Rigor in Greek Science," *Nature* 180 (October 26, 1957): 844.

29   *Thales founded abstract geometry*   George Sarton, *Introduction to the History of Science*, vol. 1: *From Homer to Omar Khayyam* (Baltimore: Williams and Wilkins, 1927), 71; also Torretti, *Philosophy of Geometry from Riemann to Poincaré*, 2.

29   *Little is known about Thales with certainty*   D. R. Dicks, "Thales," *Classical Quarterly* 9 (November 1959): 294–309; also Buchanan, "The Development of Elementary Geometry," 9–18.

32   *lifted young Pythagoras aloft*   Guthrie, *The Pythagorean Sourcebook and Library*, 61.

33   *The oracle promised Mnesarchus*   Ibid, p. 58.

33   *some of the greatest teachers of his day*   Ibid, p. 59.

33   *unable to live under the local tyrant Polycrates*   George Bruce Halsted, "The Message of Non-Euclidean Geometry," *Science* 19 (March 11, 1904): 402.

33   *settled in the city of Croton*   Benno Artman, *Euclid: The Creation of Mathematics* (New York: Springer, 1999), p. 51.

34   *the Pythagoreans, discovered*   Buchanan, "The Development of Elementary Geometry," 9–18; also Sarton, *Introduction to the History of Science*, vol. 1.

34   *swearing off beans*   Guthrie, *The Pythagorean Sourcebook and Library*, 71.

34   *Never was a historical figure more difficult to assess*   Kahn, *Pythagoras and the Pythagoreans*, 5.

35   *Pythagoras the wrestling coach*   Guthrie, *The Pythagorean Sourcebook and Library*, 154.

35   *Pythagoras or his disciples who invented music theory*   Richard Crocker, "Pythagorean Mathematics and Music," *Journal of Aesthetics and Art Criticism* 22 (Spring 1964): 325–335.

35   *a string stopped at half its length*   E. T. Whittaker, "Aristotle, Newton, Einstein," *Science* 98 (September 17, 1943): 250; also George Birkhoff, "The Origin, Nature and Influence of Relativity," *Scientific Monthly* 18 (March 1924): 228; and Torretti, *Philosophy of Geometry from Riemann to Poincaré*, p. 13.

35   *allowed to hear his teachings only*   Guthrie, *The Pythagorean Sourcebook and Library*, 76.

36   *mini-lessons in the form of questions and answers*   Ibid, p. 77.

36   *never being allowed into the same room*   Kahn, *Pythagoras and the Pythagoreans*, 8.

38   *severe, violent, and tyrannical*   Guthrie, *The Pythagorean Sourcebook and Library*, 116.

38   *the house of a man named Milo*   Ibid, p. 117.

38   *"When died the Pythagoreans,"*   Ibid, p. 135.

39   *forming a human bridge*   Ibid, p. 134.

39   *the most applied theorem in all of mathematics*   Halsted, "The Message of Non-Euclidean Geometry," 403.

39   *sacrificed one hundred oxen*   G. A. Miller, "Twenty-Five Important Topics in the History of Secondary Mathematics," *Science* 48 (August 23, 1918): 182–184.

40   *"Pythagoras transformed mathematical philosophy"*   Artman, *Euclid: The Creation of Mathematics*, 12.

40   *book seven of the* Elements   Marvin Jay Greenberg, *Euclidean and Non-Euclidean Geometries: Development and History*, 3rd ed. (New York: W. H. Freeman, 1993), 8.

41   *shifted from southern Italy to Greece*   Buchanan, "The Development of Elementary Geometry," 12.

41   *Geometry flourished in Greece*   Sarton, *Introduction to the History of Science*, vol. 1, 86, 96, 106, 111, 116, and 140; also Artman, *Euclid: The Creation of Mathematics*, 11.

43   *numerous masterworks existed*   Buchanan, "The Development of Elementary Geometry," 14.

43   *Alexander even included some scientific research*   Sarton, *Introduction to the History of Science*, vol. 1, 128.

3 The Mystery Maker

The chapter epigraph is taken from "Science News," *Science* 77 (June 2, 1933): 9a.

45   *"sandaled gentry"*   Michael Rapport, *Nineteenth-Century Europe* (New York: Palgrave Macmillan, 2005), 2.

46   *Gauss saw in Farkas a kindred spirit*   G. Waldo Dunnington, "The Sesquicentennial of the Birth of Gauss," *Scientific Monthly* 24 (May 1927): 403.

46   *"invisible passion for mathematics"*   W. K. Bühler, *Gauss: A Biographical Study* (Berlin: Springer-Verlag, 1981), 16.

46   *"I drove this problem to the point which it robbed my rest"*   George Bruce Halsted, "Biography: Bolyai Farkas. [Wolfgang Bolyai]," *American Mathematical Monthly* 3 (January 1896): 5.

47   *he founded the city of Alexandria*   Ernest W. Brown, "The History of Mathematics." *Scientific Monthly*, 12 (May, 1921): 390; also J. McKeen Cattell, "The Academy of Sciences," *Science*, 16 (December 19, 1902): 965; R. E. Langer, "Alexandria—Shrine of Mathematics," *American Mathematical Monthly* 48 (February 1941): 109; George Sarton, *Introduction to the History of Science*, vol. 1 (Baltimore: Williams and Wilkins, 1927), 149.

47   Ptolemy Soter   "The Influence of Greece on Science and Medicine," *Scientific Monthly* 3 (July 1916): 52; also Langer, "Alexandria—Shrine of Mathematics," 115; and Cattell, "The Academy of Sciences," 966.

48   Plato's academy   András Prékopa and Emil Molnár, *Non-Euclidean Geometries* (New York: Springer, 2006), 21.

48   *"Let no one unacquainted with geometry enter here"*   H. E. Buchanan, "The Development of Elementary Geometry," *Mathematics News Letter* 3 (January 1929): 12.

48   *Plato may have been inspired by the Pythagoreans*   Charles Kahn, *Pythagoras and the Pythagoreans: A Brief History* (Indianapolis: Hackett), 8.

49   *Eratosthenes, the historian and philosopher*   Sarton, *Introduction to the History of Science*, vol. 1, 172.

50   *Philosophically the book was even more profound*   George Bruce Halsted, "The Message of Non-Euclidean Geometry," *Science* 19 (March 11, 1904): 403; also P. H. Daus, "The Founding of Non-Euclidean Geometry," *Mathematics News Letter* 7 (April–May 1933): 12.

50   Theaetetus   J. J. O'Connor and E. F. Robertson, "Euclid of Alexandria," from the Web site of the School of Mathematics and Statistics at University of St. Andrews, Scotland, www-history.mcs.st-andrews.ac.uk/Biographies/Euclid.html, last accessed July 2007.

50   *None of the other books called* Elements *survived*   Jeremy Gray, *Janos Bolyai, Non-Euclidean Geometry, and the Nature of Space* (Cambridge: MIT Press, 2004), 7.

50   *To know geometry was to know the* Elements   R. L. Wilder, "The Role of Intuition," *Science* 156 (May 5, 1967): 606.

51   *The fourteenth book of Euclid*   Sarton, *Introduction to the History of Science*, vol. 1, 181.

51   *fifteenth book of Euclid*   Ibid., 427.

52   *if two lines are drawn that intersect a third*   Eric Weisstein "Euclid's Postulates," http://mathworld.wolfram.com/EuclidsPostulates.html, last accessed June 12, 2008.

52   *Following this morsel are 465 theorems*   Richard Trudeau, *The Non-Euclidean Revolution* (Boston: Birkhäuser, 1987), 5.

52   *if the converse is not self-evident*   Florence Lewis, "History of the Parallel Postulate," *American Mathematical Monthly* 27 (January 1920): 16.

52   *the scandal of elementary geometry*   Trudeau, *The Non-Euclidean Revolution*, 154.

52   *eleventh, twelfth, and thirteenth axioms*   George Bruce Halsted, "Non-Euclidean Geometry," *American Mathematical Monthly* 7 (May 1900): 124.

52   *The word "parallel"*   D. M. Y. Sommerville, *The Elements of Non-Euclidean Geometry* (Mineola, NY: Dover, 2005), 10.

53   *two lines drawn to infinity*   Lewis, "History of the Parallel Postulate," 17; also Roberto Bonolo, *Non-Euclidean Geometry* (New York: Dover, 1955); and Robert David Sack, "Geography, Geometry, and Explanation," *Annals of the Association of American Geographers* 62 (March 1972): 62.

53   *They did not doubt for a second*   Lewis, "History of the Parallel Postulate," 16; also Gray, *Janos Bolyai, Non-Euclidean Geometry, and the Nature of Space*, 13.

54   *The postulates are the foundation blocks*   Warren Weaver, "Scientific Explanation," *Science* 143 (March 20, 1964): 1298.

54   *Without it, much of the first book of* Elements *could not be written*   Gray, *Janos Bolyai, Non-Euclidean Geometry, and the Nature of Space*, 19.

54   *the fifth postulate to be true*   V. Kagan, *N. Lobachevsky: His Contribution to Science* (Moscow: Foreign Language Publishing House, 1957), 18; also Lewis, "History of the Parallel Postulate," 17.

54   *Every generation of mathematicians after Euclid*   Alexander Ziwet, "Review of 'Nicolái Ivánovich Lobachévsky' by A. Vasiliev," *Science* 1 (March 29, 1895): 356–358; also Judith V. Grabiner, "The Centrality of Mathematics in the History of Western Thought," *Mathematics Magazine* 61 (October 1988): 221.

55   *the "knowledge of that which always is"*   Grabiner, "The Centrality of Mathematics in the History of Western Thought," 220.

4 Those False and Would-Be Proofs

The chapter epigraph is taken from Florence Lewis, "History of the Parallel Postulate," *American Mathematical Monthly* 27 (January 1920): 16. I chose to change the wording from "the" to "[an]" to avoid readers mistakenly thinking that Aristotle was referring specifically to the fifth postulate.

57   *Robert Recorde*   J. J. O'Connor and E. F. Robertson, "Robert Recorde," from the Web site of the School of Mathematics and Statistics at the University of St. Andrews, Scotland, www.history.mcs.st-andrews.ac.uk/Biographies/Recorde.html, last accessed July 2007.

57   *Euclid did not include this postulate among the originals*   Florence Lewis, "History of the Parallel Postulate," *American Mathematical Monthly* 27 (January 1920): 16.

58   *Posidonius thought he could prove*   Richard Trudeau, *The Non-Euclidean Revolution* (Boston: Birkhäuser, 1987), 119.

58   Geminus of Rhodes   Lewis, "History of the Parallel Postulate," 17; also George Bruce Halsted, "Non-Euclidean Geometry," *American Mathematical Monthly* 7 (May 1900): 124; and J. J. O'Connor and E. F. Robertson, "Geminus," from the Web site of the School of Mathematics and Statistics at the University of St. Andrews, Scotland, www.history.mcs.st-andrews.ac.uk/Biographies/Geminus .html, last accessed July 2007.

59   *they contributed little to the advancement of the discipline*   Ernest W. Brown, "The History of Mathematics," *Scientific Monthly* 12 (May 1921): 390; also Max Dehn, "Mathematics, 200 BC–600 AD," *American Mathematical Monthly* 51 (March 1944): 156; George Sarton, *Introduction to the History of Science*, vol. 1: *From Homer to Omar Khayyam* (Baltimore: Williams and Wilkins, 1927), 267; and William Salant, "Science and Society in Ancient Rome," *Scientific Monthly* 47 (December 1938): p. 525.

59   *The Greeks achieved more than that*   Sarton, *Introduction to the History of Science*, vol. 1, 109, 201.

60   Hero of Alexandria   Salant, "Science and Society in Ancient Rome," p. 528.

60   Theon of Alexandria   Sarton, *Introduction to the History of Science*, vol. 1, 367; also R. E. Langer, "Alexandria—Shrine of Mathematics," *American Mathematical Monthly* 48 (February 1941): 124.

60   *numerous corrections*   Trudeau, *The Non-Euclidean Revolution*, 22; also J. J. O'Connor and E. F. Robertson, "Theon of Alexandria," from the Web site of the School of Mathematics and Statistics at the University of St. Andrews, Scotland, www.history.mcs.st-andrews. ac.uk/Biographies/Theon.html, last accessed July 2007. My own version is not based on Campanus's translation of Theon's version but on an earlier Greek text discovered in the nineteenth century and translated into English by Sir Thomas Heath.

60   *suffered a terrible fate*   Sarton, *Introduction to the History of Science*, vol. 1, 387.

60   Proclus   Ibid., 402; also Lewis, "History of the Parallel Postulate," 17; and Roberto Bonolo, *Non-Euclidean Geometry* (New York: Dover, 1955), 4.

61   *"ought to be struck from the postulates altogether"*   Roberto Torretti, *Philosophy of Geometry from Riemann to Poincaré* (Dordrecht, Holland: D. Reidel Publishing Co., 1984), 41.

61   *Justinian closed the Athens Academy*   Sarton, *Introduction to the History of Science*, vol. 1, 414.

61   Khalid ibn Yazid   Ibid., 495; Sarton also discusses the important work in mathematics done by Islamic scholars for several hundred years beginning in the mid-eighth century.

62   Abdallah al-Ma'mon   Ibid., 557.

62   the Ibn Musa brothers   Ibid., 560.

62   al-Hajjaj ibn Yusuf   Ibid., 562.

63   Hunain ibn Ishaq   J. J. O'Connor and E. F. Robertson, "Al-Sabi Thabit ibn Qurra al-Harrani," from the Web site of the School of Mathematics and Statistics at the University of St. Andrews, Scotland, www.history.mcs.st-andrews.ac.uk/ Biographies/Thabit.html, last accessed July 2007; also O'Connor and Robertson, "Abu Zayd Hunayn ibn Ishaq al-Ibadi," ibid.; and Sarton, *Introduction to the History of Science*, vol. 1, 611.

63   Thabit ibn Qurra   Sarton, *Introduction to the History of Science*, vol. 1, 599.

63   al-Khwarizmi   Ibid., p., 563.

64   Gerbert   Ibid., p. 669.

64   *In Moorish Spain*   Ibid., pp. 658, 715.

64   Beruelius   Ibid., p. 695.

65   Nassiruddin at-Tusi   J. J. O'Connor and E. F. Robertson, "al-Ishbili Abu Mu-hammad Jabir ibn Aflah," from the Web site of the School of Mathematics and Statistics at the University of St. Andrews, Scotland, www.history.mcs.st-andrews.ac.uk/Biographies/Jabir_ibn_Aflah.html, last accessed July 2007; J. Stephenson, "The Classification of the Sciences according to Nasiruddin Tusi," *Isis* 5 (1923): 329–338; Bonolo, *Non-Euclidean Geometry*, 10; V. Kagan, *N. Lobachevsky: His Contribution to Science* (Moscow: Foreign Language Publishing House, 1957), 21; and András Prékopa and Emil Molnár, *Non-Euclidean Geometries* (New York: Springer, 2006), 23.

65   Omar Khayyam   Sarton, *Introduction to the History of Science*, vol. 1, 738.

65   *proving the fifth postulate*   Lewis, "History of the Parallel Postulate," 18; also George Bruce Halsted, "Bibliography of Hyper-Space and Non-Euclidean Geometry," *American Journal of Mathematics* 1 (1878): 261–276; and Halsted, "Non-Euclidean Geometry," 125.

66   *in those days before the printing press*   Bonolo, *Non-Euclidean Geometry*, 12.

66   *the abacus*   Sarton, *Introduction to the History of Science*, vol. 1, 740.

66   Adelard   Charles H. Haskins, "Arabic Science in Western Europe," *Isis* 7 (1925): 497; also Haskins, "Adelard of Bath," *English Historical Review* 26 (July 1911): 491–498; and J. J. O'Connor and E. F. Robertson, "Adelard of Bath," from the Web site of the School of Mathematics and Statistics at the University of St. Andrews, Scotland, www.history.mcs.st-andrews.ac.uk/Biographies/Adelard.html, last accessed July 2007.

67   *Adelard was not alone*   Scott L. Montgomery, "Gained in the Translation—Scientific Knowledge Is Enriched as It Moves between Languages," *Nature* 409 (February 8, 2001): 667; also Haskins, "Arabic Science in Western Europe," 478; and Haskins, "Adelard of Bath," 491–498.

67   *Universities in Europe*   Brown, "The History of Mathematics," 385.

67   *By the end of the thirteenth century*   Florian Cajori, "Attempts Made During the Eighteenth and Nineteenth Centuries to Reform the Teaching of Geometry," *American Mathematical Monthly* 17 (October 1910): 192.

67   Leonardo of Pisa   Walter C. Eells, "Discussions: The Ten Most Important Mathematical Books in the World," *American Mathematical Monthly* 30 (September–October 1923): 319.

68   Campanus of Novara   C. S. Peirce, "Campanus," *Science* 13 (May 24, 1901): 809–811; also J. J. O'Connor and E. F. Robertson, "Campanus of Novara," from the Web site of the School of Mathematics and Statistics at the University of St. Andrews, Scotland, www.history.mcs.st-andrews.ac.uk/Biographies/Campanus.html, last accessed July 2007.

68   Levi ben Gerson   J. J. O'Connor and E. F. Robertson, "Levi ben Gerson," from the Web site of the School of Mathematics and Statistics at the University of St. Andrews, Scotland, www.history.mcs.st-andrews.ac.uk/Biographies/Levi.html, last accessed July 2007.

68   Sir Henry Billingsley   George Bruce Halsted, "The Message of Non-Euclidean Geometry," *Science* 19 (March 11, 1904): 403.

68   Pietro Antonio Cataldi   Lewis, "History of the Parallel Postulate," 18; also Trudeau, *The Non-Euclidean Revolution*, 128; and Bonolo, *Non-Euclidean Geometry*, 13.

68   Claudius Ptolemy   Sarton, *Introduction to the History of Science*, vol. 1, 273; also Lewis, "History of the Parallel Postulate," 17.

69   *"The cosmogonist had to fill the skies"* Arthur Stanley Eddington, "The Theory of Relativity and Its Influence on Scientific Thought," *Scientific Monthly* 16 (January 1923): 34.

69   astronomy in the seventeenth century Jeremy Gray, *János Bolyai: Non-Euclidean Geometry and the Nature of Space* (Cambridge, MA: Burndy Library, 2004): 21; also Torretti, *Philosophy of Geometry from Riemann to Poincaré*, 23.

69   *likened to a violent act* Shiing-Shen Chern, "From Triangles to Manifolds," *American Mathematical Monthly* 86 (May 1979): 343.

69   *"There cannot be a language more universal and more simple"* Fourier quoted in Walter C. Eells, "Discussions: The Ten Most Important Mathematical Books in the World," *American Mathematical Monthly* 30 (September–October 1923): 320.

5 A Codebreaker's Fix

The chapter epigraph is taken from Philip Beeley and Christopher J. Scriba, *The Correspondence of John Wallis: Volume I (1641–1659)* (Oxford: Oxford University Press, 2003), xlii.

72   *two hours' time* Ibid., p. xxvii.

72   *decoded enough of King Louis XIV* Ibid, p. xix.

73   *In recognition of Wallis's efforts* Jacqueline A. Stedall, *The Arithmetic of Infinitesimals: John Wallis 1656* (New York: Springer-Verlag, 2004), p. xii.

73   Sir Henry Savile Beeley and Scriba, *The Correspondence of John Wallis*, xxxv.

73   Peter Turner J. F. Scott, *The Mathematical work of John Wallis* (New York: Chelsea Publishing Co., 2004), 14.

73   *deaf-mutes* Beeley and Scriba, *The Correspondence of John Wallis*, xvii.

74   *defined his career* Stedall, *The Arithmetic of Infinitesimals*, xi.

74   *struggling to emerge* Ibid, p. xii.

74   *influenced both Newton and Leibniz* Beeley and Scriba, *The Correspondence of John Wallis*, xvii.

74   David Gregory Stedall, *The Arithmetic if Infinitesimals*, xxxiii.

74   *avoid the fickle fate* Beeley and Scriba, *The Correspondence of John Wallis*, xviii.

75   *enamored of the ancient thinkers* Ibid, p. xviii.

75   *Wallis's attempt* D. M. Y. Sommerville, *The Elements of Non-Euclidean Geometry* (Mineola, NY: Dover, 2005), 7.

76   *It was not a proof, however* Roberto Bonolo, *Non-Euclidean Geometry* (New York: Dover, 1955), 16.

76   *Legendre tried and failed* George Bruce Halsted, "Non-Euclidean Geometry," *American Mathematical Monthly* 7 (May 1900): 125.

76   *new editions of Euclid's book* Alexander Ziwet, "Euclid as a Text-Book of Geometry," *Science* 4 (November 7, 1884): 442.

77   *surveying, billiards, warfare* Florian Cajori, "Attempts Made During the Eighteenth and Nineteenth Centuries to Reform the Teaching of Geometry," *American Mathematical Monthly* 17 (October 1910): 182.

77   *why anyone would bother with Euclid* Roberto Torretti, "Nineteenth Century Geometry (2003 revision)," *Stanford Encyclopedia of Philosophy*, http://plato.stanford.edu/entries/geometry-19th/

77   *the only country in Europe where Euclid was still taught* Cajori, "Attempts Made During the Eighteenth and Nineteenth Centuries to Reform the Teaching of Geometry," 195.

77   *Legendre's 1794 book* Abe Shenitzer, "How Hyperbolic Geometry Became Respectable," *American Mathematical Monthly* 101 (May 1994): 465; also Cajori, "Attempts Made During the Eighteenth and Nineteenth Centuries to Reform the Teaching of Geometry," 185.

78   *He again resorted to the tried-and-untrue proof* George Bruce Halsted, "Report on Progress in Non-Euclidean Geometry," *Science* 10 (October 20, 1899): 555; also Bonolo, *Non-Euclidean Geometry*, 60.

78   *the Reformed College of Maros-Vásárhely* Halsted, "Biography: Bolyai Farkas. [Wolfgang Bolyai]," *American Mathematical Monthly* 3 (January 1896), 2.

78   *"That feeling of seeing each other for the last time is indescribable"* W. K. Bühler, *Gauss: A Biographical Study* (Berlin: Springer-Verlag, 1981), 17.

78   *He shared some of this work with Gauss* Roberto Torretti, *Philosophy of Geometry from Riemann to Poincaré* (Dordrecht, Holland: D. Reidel Publishing Co., 1984), 54; also George Bruce Halsted, "The Foundations of Geometry," *Science* 6 (September 24, 1897): 488.

79   *"As for me, I have already made some progress in my work"* Bonolo, *Non-Euclidean Geometry*, 65.

79   *After he became a professor of mathematics* G. Waldo Dunnington, *Gauss: Titan of Science* (Washington, DC: Mathematics Association of America, 2004), 177; also Halsted, "The Foundations of Geometry," 489.

80   *"I still hope that these cliffs"* Gauss quoted in Dunnington, *Gauss: Titan of Science*, 100.

80   *the fundamental theory of algebra* G. Waldo Dunnington, "The Historical Significance of Carl Friedrich Gauss in Mathematics and Some Aspects of His Work," *Mathematics News Letter* 8 (May 1934): 178.

80   *"Doctoral dissertations, even of the greatest scholars"* Dunnington, *Gauss: Titan of Science*, 40.

81   *the* Disquisitiones Arithmeticae   Ibid.; also Bühler, *Gauss: A Biographical Study*, 32.

81   *Gauss published only seven* G. E. Burch, "Carl Friedrich Gauss—a Genius Who Apparently Died of Arteriosclerotic Heart Disease and Congestive Heart Failure," *A.M.A. Archives of Internal Medicine* 101 (April 1958): 827.

81   *No mistakes have ever been found in the book* Dunnington, *Gauss: Titan of Science*, 40.

82   *"Your* Disquisitiones *have with one stroke elevated you"* Lagrange quoted in ibid., p. 43.

82   *"a supernatural spirit in a human body"* Teets and Whitehead, "The Discovery of Ceres," 83.

82   *wise and liberal support* Ibid., p. 41.

82   *misspelling his name* Catherine Goldstein et al., *The Shaping of Arithmetic after C. F. Gauss's Disquisitiones Arithmeticae* (Berlin: Springer-Verlag, 2007), 19.

6 Searching for Ceres

83   Giuseppe Piazzi C. L. Doolittle, "Some Advances Made in Astronomical Science during the Nineteenth Century," *Science* 14 (July 5, 1901): 1–12.

83   *a comet without a tail* G. Waldo Dunnington, *Gauss: Titan of Science* (Washington, DC: Mathematical Association of America, 2004), 49.

83   *"something better than a comet"* Richard P. Binzel, "Asteroids Come of Age," *Science* 289 (September 22, 2000): 2065.

84   Titus and Bode   C. T. Russell et al., "DAWN: A Journey to the Beginning of the Solar System," ACM Conference Paper, www.ssc.igpp.ucla.edu/dawn/pdf/ACMConferencePaper, last accessed May 2007.

84   *He held that different strings gave different notes*   George Birkhoff, "The Origin, Nature and Influence of Relativity," *Scientific Monthly* 18 (March 1924): 228.

84   Baron von Zach   Donald A. Teets and Karen Whitehead, "The Discovery of Ceres: How Gauss Became Famous," *Mathematics Magazine* 72 (April 1999): 83; also W. K. Bühler, *Gauss: A Biographical Study* (Berlin: Springer-Verlag, 1981), 43.

87   *"the noblest record of its intelligence"*   D. W. Morehouse, "Astronomy's Contribution to the Stream of Human Thought," *Science* 75 (January 8, 1932): 27.

87   *"I abjure, curse, and detest the said errors"*   Otto Struve, "Freedom of Thought in Astronomy," *Scientific Monthly* 40 (March 1935): 250.

87   *The intricate and irregular motion of the planets*   Edgar W. Woodard, "The Calculation of Planetary Motions," *National Mathematics Magazine* 14 (January 1940): 179.

87   celestial mechanics   Doolittle, "Some Advances Made in Astronomical Science during the Nineteenth Century," 2. The problem of calculating the orbit of planets versus comets is also described in Woodard, "The Calculation of Planetary Motions," 183.

87   *the problem of determining an orbit*   Dunnington, *Gauss: Titan of Science*, 53. A thorough description of the method for computing planetary orbits can be found in Teets and Whitehead, "The Computation of Planetary Orbits," 397–404.

88   *With only a few observations*   R. L. Plackett, "Studies in the History of Probability and Statistics. XXIX: The Discovery of the Method of Least Squares," *Biometrika* 59 (August 1972): 240.

89   *"Could I ever have found a more seasonable opportunity"*   Teets and Whitehead, "The Discovery of Ceres," p. 84.

89   Heinrich Olbers   Dunnington, *Gauss: Titan of Science*, 56.

90   *the method of least squares*   Ibid., p. 54; also Bühler, *Gauss: A Biographical Study*, 84.

90   *the single most valuable tool ever devised for analyzing data*   R. S. Woodward, "The Century's Progress in Applied Mathematics," *Science* 11 (January 19, 1900): 83.

91   *"Few branches of science owe so large a proportion"*   Dunnington, *Gauss: Titan of Science*, 20.

91   *When the mathematician Abel later published*   Bühler, *Gauss: A Biographical Study*, 86.

92   *"magnetic poles towards which the compass of my mind ever turns"*   Ibid., p. 61.

92   *He also named his firstborn son Joseph*   Ibid., p. 224; Dunnington also mentions that Gauss named several of his other children after the discoverers of other asteroids.

92   *The success launched Gauss's career as well*   Truman Henry Safford, "Astronomy in the First Half of the Nineteenth Century," *Science* 10 (December 29, 1899): 962–963; also Dunnington, *Gauss: Titan of Science*, 58, 251.

92   *The Royal Society of London elected him a foreign member in 1804*   B. F. Finkel, "Biography: Karl Frederich Gauss," *American Mathematical Monthly* 8 (February 1901): 29; and S. R. C. Malin, "Sesquicentenary of Gauss's First Measurement of the Absolute Value of Magnetic Intensity," *Philosophical Transactions of the Royal Society of London, Series A, Mathematical and Physical Sciences* 306, no. 1492, The Earth's Core: Its Structure, Evolution and Magnetic Field (August 20, 1982): 5–8.

92   *The Academy of Sciences at St. Petersburg*   G. Waldo Dunnington, "The Sesquicentennial of the Birth of Gauss," *Scientific Monthly* 24 (May 1927): 403.

93   *Gauss subsequently set about writing down*   Bühler, *Gauss: A Biographical Study*, 86.
93   *Legendre was not satisfied*   Plackett, "Studies in the History of Probability and Statistics," 245.
94   *"It is scarcely comprehensible how men of honor"*   Dunnington, *Gauss: Titan of Science*, 73.

7 The Dim Light of Exhaustion

The chapter epigraph is taken from V. N. Kagan, *Lobachevsky: His Contribution to Science* (Moscow: Foreign Language Publishing House, 1957), 40.

95   *Gauss was always cautious*   G. Waldo Dunnington, *Gauss: Titan of Science* (Washington, DC: Mathematical Association of America, 2004), 214, 176.
95   *He wrote to his friend Heinrich Olbers*   W. K. Bühler, *Gauss: A Biographical Study* (Berlin: Springer-Verlag, 1981), 35.
96   *"somewhat dull feelers"*   Dunnington, *Gauss: Titan of Science*, 215.
96   *"few but ripe"*   Ibid., p. 72.
97   *a tanner named Mr. Ritter*   Dunnington, *Gauss: Titan of Science*, 61.
97   Johanna Elisabeth Rosina Osthoff   Ibid., p. 48; also G. Waldo Dunnington, "The Sesquicentennial of the Birth of Gauss," *Scientific Monthly* 24 (May 1927): 403; and Dunnington, *Gauss: Titan of Science*, 62.
97   *"exactly such a girl"*   Dunnington, *Gauss: Titan of Science*, 62.
98   *"The white snow passes away"*   Ibid., p. 63.
98   *an almost unquenchable thirst for power*   Michael Rapport, *Nineteenth-Century Europe* (New York: Palgrave Macmillan, 2005), 35.
99   *a mournful, slow wagon train*   Dunnington, *Gauss: Titan of Science*, 80; also Bühler, *Gauss: A Biographical Study*, 54; and Rapport, *Nineteenth-Century Europe*, 36.
99   *Confederation of the Rhine*   Rapport, *Nineteenth-Century Europe*, 36.
100 Disquisitiones   Dunnington, *Gauss: Titan of Science*, 67.
100 Sophie Germain   Jesse A. Fernandez Martinez, "Sophie Germain," *Scientific Monthly* 63 (October 1946): 257–260.
101 *"The foremost mathematician of his time lives there"*   Dunnington, *Gauss: Titan of Science*, 44.
101 *was planning to sell Göttingen*   Ibid., p. 64.
102 *French occupation brought a few good things*   Rapport, *Nineteenth-Century Europe*, 45.
103 *as early as 1792*   George Bruce Halsted, "Simon's Claim for Gauss in Non-Euclidean Geometry," *American Mathematical Monthly* 11 (April 1904): 85.
103 *Farkas Bolyai had a breakthrough*   Ibid., p. 85.
103 *an experimental science*   Richard Trudeau, *The Non-Euclidean Revolution* (Boston: Birkhäuser, 1987), 148.

8 Gauss's Little Secret

The chapter epigraph is taken from George Bruce Halsted, "Biography: Felix Klein," *American Mathematical Monthly* 1 (December 1894): 419.

105 Giovanni Girolamo Saccheri   J. R. Lucas, "Euclides ab omni naevo vindicatus" [Euclid freed from any blemish], *British Journal for the Philosophy of Science* 20 (May 1969): 3; also George Bruce Halsted, "A Class-Book of Non-Euclidean Geometry," *American Mathematical Monthly* 8 (April 1901): 87; Halsted,

"Non-Euclidean Geometry," *American Mathematical Monthly* 7 (May 1900): 126; and Halsted, "Non-Euclidean Geometry: Historical and Expository," *American Mathematical Monthly* 1 (August 1894): 151.

107   Georg Simon Klügel   Florence Lewis, "History of the Parallel Postulate," *American Mathematical Monthly* 27 (January 1920): 18; also Roberto Bonolo, *Non-Euclidean Geometry* (New York: Dover, 1955), 44.

108   John Henry Lambert   Cornelius Lanczos, *Space through the Ages: The Evolution of Geometrical Ideas from Pythagoras to Hilbert and Einstein* (New York: Academic Press, 1970), 63; also V. Kagan, *N. Lobachevsky: His Contribution to Science* (Moscow: Foreign Language Publishing House, 1957), 22.

109   *He wrote to his student*   Roberto Torretti, *Philosophy of Geometry from Riemann to Poincaré* (Dordrecht, Holland: D. Reidel Publishing Co., 1984), 55.

109   *"Perhaps only in another life"*   Jeremy Gray, *Janos Bolyai, Non-Euclidean Geometry, and the Nature of Space* (Cambridge: MIT Press, 2004), 44.

110   *Gauss read the* Critique *five times*   G. Waldo Dunnington, *Gauss: Titan of Science* (Washington, DC: Mathematical Association of America, 2004), 315.

110   *the "clamor of the Boetians"*   P. H. Daus, "The Founding of Non-Euclidean Geometry," *Mathematics News Letter* 7 (April–May 1933): 14; also Dunnington, *Gauss: Titan of Science*, 182; and Lanczos, *Space through the Ages*, 65.

9 Lessons of Curvature

The chapter epigraph is taken from Martin Jay Greenberg, *Euclidean and Non-Euclidean Geometries: Development and History, 3rd ed.* (New York: W.H. Freeman, 1993), 5.

111   Czar Alexander I   András Prékopa and Emil Molnár, *Non-Euclidean Geometries* (New York: Springer, 2006), 9.

111   *"The teaching of physics and mathematics in Kazan"*   V. Kagan, *N. Lobachevsky: His Contribution to Science* (Moscow: Foreign Language Publishing House, 1957), 26.

111   Johann Christian Martin Bartels   Alexander Vucinich, "Nikolai Ivanovich Lobachevskii: The Man behind the First Non-Euclidean Geometry," *Isis* 53 (December 1962): 470.

112   *Lobachevsky was born in 1792*   Ibid., 465–481. Lobachevsky's birth year is sometimes reported as 1793.

112   *Lobachevsky's boyhood*   E. T. Bell, *Men of Mathematics* (New York: Simon and Schuster, 1986), 294; also George Bruce Halsted, "The Message of Non-Euclidean Geometry," *Science* 19 (March 11, 1904): 404.

113   *the young János Bolyai*   Halsted, "The Message of Non-Euclidean Geometry," 408.

113   *Farkas and his wife were unhappy*   Prékopa and Molnár, *Non-Euclidean Geometries*, 2; also Jeremy Gray, *Janos Bolyai, Non-Euclidean Geometry, and the Nature of Space* (Cambridge: MIT Press, 2004), 55.

113   *János mastered geometry and calculus within a few years*   Roberto Bonolo, *Non-Euclidean Geometry* (New York: Dover, 1955), 96.

114   *a picture of Gauss on his wall*   Gray, *Janos Bolyai, Non-Euclidean Geometry, and the Nature of Space*, 49.

114   *the greatest Hungarian mathematician of all time*   Ibid., 3.

114   *"I closed her angelic eyes"*   G. Waldo Dunnington, *Gauss: Titan of Science* (Washington, DC: Mathematical Association of America, 2004), 93.

115   Minna Gauss   W. K. Bühler, *Gauss: A Biographical Study* (Berlin: Springer-Verlag, 1981), 67; also G. Waldo Dunnington, "The Sesquicentennial of the Birth of Gauss," *Scientific Monthly* 24 (May 1927): 405.

115   *"I can offer you only a divided heart"*   Bühler, *Gauss: A Biographical Study*, 67.

116   *Lobachevsky's commandant complained to the rector*   Halsted, "The Message of Non-Euclidean Geometry," 408; also Alexander Vucinich, "Nikolai Ivanovich Lobachevskii, 470; and Kagan, *N. Lobachevsky: His Contribution to Science*, 29.

117   *the foreign professors were expelled*   Prékopa and Molnár, *Non-Euclidean Geometries*, 9.

117   *In the midst of all this*   D. M. Y. Sommerville, *The Elements of Non-Euclidean Geometry* (Mineola, NY: Dover, 2005), 20.

118   *how much his thinking had evolved*   George Bruce Halsted, "Gauss and the Non-Euclidean Geometry," *American Mathematical Monthly* 7 (November 1900): 250; also Dunnington, *Gauss: Titan of Science*, 117.

118   *the backbone of physics*   Ibid., p. 140.

118   *The basis of applied mathematics*   Dorothy L. Bernstein, "The Role of Applications in Pure Mathematics," *American Mathematical Monthly* 86 (April 1979): 246.

119   *no publishing of ideas that were not fully formed*   Dunnington, *Gauss: Titan of Science*, 209; also Ian Stewart, "Justifying the Means," *Nature* 354 (November 21, 1991): 184–186.

119   *"This person had the impudence"*   Raymond C. Archibald, "Gauss and the Regular Polygon of Seventeen Sides," *American Mathematical Monthly* 27 (July–September 1920): 323–326.

120   *frustrating failure*   Dunnington, *Gauss: Titan of Science*, 178.

121   *"vain effort to conceal with an untenable tissue"*   George Bruce Halsted, "Report on Progress in Non-Euclidean Geometry," *Science* 10 (October 20, 1899): 556.

121   *Farkas seems to have fallen into some sort of dark misogynistic space*   Prékopa and Molnár, *Non-Euclidean Geometries*, 15.

10 To Stir the Nests of Wasps

The chapter epigraph is taken from *La Geometrie* as translated in G. A. Miller, "General or Special in the Development of Mathematics," *Science* 93 (March 7, 1941): 235.

123   Gauss's mother   G. Waldo Dunnington, *Gauss: Titan of Science* (Washington, DC: Mathematical Association of America, 2004), 140, 203.

124   *he had worked intermittently on non-Euclidean geometry*   Ibid., p. 180.

125   *a severe critic of others*   W. K. Bühler, *Gauss: A Biographical Study* (Berlin: Springer-Verlag, 1981), 79.

125   *the language of his first book was most refined*   Ibid., p. 80.

125   Schwab and Metternich   Dunnington, *Gauss: Titan of Science*, 214.

125   Friedrich Ludwig Wachter   Ibid., p. 180; also Roberto Bonolo, *Non-Euclidean Geometry* (New York: Dover, 1955), 62.

125   *sidetracked by the Napoleonic Wars*   Dunnington, *Gauss: Titan of Science*, 268.

126   *In this letter, Wachter speculated*   George Bruce Halsted, "Gauss and the Non-Euclidean Geometry," *American Mathematical Monthly* 7 (November 1900): 250.

126   *"anti-Euclidean" geometry*   Bonolo, *Non-Euclidean Geometry*, 63; also Halsted, "Gauss and the Non-Euclidean Geometry." *American Mathematical Monthly* 7 (November 1900): 249.

127   *"astral" geometry*   George Bruce Halsted, "Biography: Bolyai Farkas. [Wolfgang Bolyai]," *American Mathematical Monthly* 3 (January 1896): 4; also Halsted, "Gauss and the Non-Euclidean Geometry," 250.

127   *Gauss was thrilled*   Jeremy Gray, *Janos Bolyai, Non-Euclidean Geometry, and the Nature of Space* (Cambridge: MIT Press, 2004), 48; also Roberto Torretti, *Philosophy of Geometry from Riemann to Poincaré* (Dordrecht, Holland: D. Reidel Publishing Co., 1984), 52.

128   *Schweikart had invented his astral geometry*   George Bruce Halsted, "A Class-Book of Non-Euclidean Geometry," *American Mathematical Monthly* 8 (April 1901): 87.

128   *Franz Adolof Taurinus*   Halsted, "Gauss and the Non-Euclidean Geometry" 252; also Torretti, *Philosophy of Geometry from Riemann to Poincaré*, 52.

128   *in a letter on November 8, 1824*   D. M. Y. Sommerville, *The Elements of Non-Euclidean Geometry* (Mineola, NY: Dover, 2005), 14; and Torretti, *Philosophy of Geometry from Riemann to Poincaré*, 55.

128   *He published a book in 1825*   V. Kagan, *N. Lobachevsky: His Contribution to Science* (Moscow: Foreign Language Publishing House, 1957), 25.

129   *which his notebooks would later reveal*   D. J. Struik, "Outline of a History of Differential Geometry II," *Isis* 20 (November 1933): 164.

130   *the heliotrope*   Dunnington, *Gauss: Titan of Science*, 213; also Dunnington, "The Sesquicentennial of the Birth of Gauss," *Scientific Monthly* 24 (May 1927): 409.

130   *This work was important to the government*   Carl Friedrich Gauss, *General Investigations of Curved Surfaces*, ed. Peter Pesic (Mineola, NY: Dover, 2005), iv.

130   *Gauss's interest in making such measurements*   Dunnington, *Gauss: Titan of Science*, 116.

131   *From a given point at a given latitude and longitude*   Bühler, *Gauss: A Biographical Study*, 95.

132   *he wasted decades*   Dunnington, *Gauss: Titan of Science*, 213.

132   *a million pieces of data*   Gauss, *General Investigations of Curved Surfaces*, vi.

133   *"clods of dirt"*   Dunnington, *Gauss: Titan of Science*, 300.

133   *Inselberg, Brocken, and Hohenhagen*   Gauss, *General Investigations of Curved Surfaces*, p. vi; also Dunnington, "The Sesquicentennial of the Birth of Gauss," 409.

133   *"All the measurements in the world"*   Dunnington, *Gauss: Titan of Science*, 312.

134   geodesy   Dunnington, *Gauss: Titan of Science*, 98.

134   *if Gauss had not invented the geometry of curved surfaces*   Ibid., p. 350.

134   conformal mapping   Its definition is taken from the NASA Web site, www.grc.nasa.gov/WWW/K-12/airplane/map.html, last accessed May 22, 2008.

134   *Berlin began seriously wooing him*   Dunnington, *Gauss: Titan of Science*, 132; also Raymond C. Archibald, "History of Mathematics After the Sixteenth Century," *Mathematics* (January 1949): 47.

135   *Minna was increasingly bedridden*   Dunnington, *Gauss: Titan of Science*, 102.

11 A Strange New World

The chapter epigraph is taken from, G. Waldo Dunnington, *Gauss: Titan of Science* (Washington, DC: Mathematics Association of America, 2004): p. 188.

137   *Farkas's salary was stretched thin*   András Prékopa and Emil Molnár, *Non-Euclidean Geometries* (New York: Springer, 2006), 15.

137   *multiple languages*   J. J. O'Connor and E. F. Robertson, "János Bolyai," from the Web site of the School of Mathematics and Statistics at the University of St. Andrews, Scotland, www.history.mcs.st-andrews.ac.uk/Biographies/Bolyai.html, last accessed July 2007.

138   Carl Szász   Roberto Bonolo, *Non-Euclidean Geometry* (New York: Dover, 1955), 97.

139   *"I have traversed this bottomless night"* Jeremy Gray, *Janos Bolyai, Non-Euclidean Geometry, and the Nature of Space* (Cambridge: MIT Press, 2004), 51.

139   *"science of absolute space"* George Bruce Halsted, "Non-Euclidean Geometry," *American Mathematical Monthly* 7 (May 1900); also Roger Osborne, "Some Historic and Philosophic Aspects of Geometry," *Mathematics Magazine* 24, no. 2 (November–December 1950): 77–82; and John Dolman Jr., "Remarks on Professor Lyle's Postulate I. of Euclid's E," *American Mathematical Monthly* 1 (April 1894): 116.

140   *Euclid's phoenix* Dunnington, *Gauss: Titan of Science*, 183.

141   *detailed and wonderfully clear* J. J. O'Connor and E. F. Robertson, "Nikolai Ivanovich Lobachevsky," from the Web site of the School of Mathematics and Statistics at the University of St. Andrews, Scotland, www.history.mcs.st-andrews.ac.uk/Biographies/Lobachevsky.html, last accessed July 2007.

142   *a great admirer of Kant* George Bruce Halsted et al., "The Appreciation of Non-Euclidean Geometry," *Science* 13 (March 22, 1901): 463; also Alexander Vucinich, "Nikolai Ivanovich Lobachevskii: The Man behind the First Non-Euclidean Geometry," *Isis* 53 (December 1962): 475.

142   *"a wretched and shameful spectacle"* V. Kagan, *N. Lobachevsky: His Contribution to Science* (Moscow: Foreign Language Publishing House, 1957), 61.

142   Geometriya Dunnington, *Gauss: Titan of Science*, 185; also Kagan, *N. Lobachevsky: His Contribution to Science*, 32; P. H. Daus, "The Founding of Non-Euclidean Geometry," *Mathematics News Letter* 7 (April–May 1933): 15; and Vucinich, "Nikolai Ivanovich Lobachevskii: The Man behind the First Non-Euclidean Geometry," 472.

142   *"No vigorous proof of this truth"* Kagan, *N. Lobachevsky: His Contribution to Science*, 33.

142   Magnitsky Vucinich, "Nikolai Ivanovich Lobachevskii: The Man behind the First Non-Euclidean Geometry," 465.

145   *the foundations of geometry* Dunnington, *Gauss: Titan of Science*, 99, 182.

145   *small beings on a huge planet* Hermann Helmholtz, "The Origin and Meaning of Geometrical Axioms," *Mind* 1 (July 1876): 305.

145   *the nature of curved surfaces* Ernest W. Brown, "The History of Mathematics," *Scientific Monthly* 12 (May 1921): 404; also Dunnington, *Gauss: Titan of Science*, 110; and Gray, *Janos Bolyai, Non-Euclidean Geometry, and the Nature of Space*, 84.

145   *the curvature remains the same* Carl Friedrich Gauss, *General Investigations of Curved Surfaces*, ed. Peter Pesic (Mineola, NY: Dover, 2005), v.

12 A Message for You, Ambassador

The chapter epigraph is taken from Euclid, *Elements* (Santa Fe: Green Lion Press, 2002), 55. Emphasis added.

147   cholera epidemic George C. Kohn, *The Wordsworth Encyclopedia of Plague and Pestilence* (Hertfordshire, UK: Wordsworth Reference, 1995), 27.

148   *Lobachevsky sought to isolate the university* V. Kagan, *N. Lobachevsky: His Contribution to Science* (Moscow: Foreign Language Publishing House, 1957), 63.

148   *A highly talented administrator* Ibid., p. 63; also Alexander Ziwet, "Review of 'Nicolái Ivánovich Lobachévsky' by A. Vasiliev," *Science* 1 (March 29, 1895): 356–358.

149   *challenged the conventional notions of space and geometry* Norman Daniels, "Lobachevsky: Some Anticipations of Later Views on the Relation between Geometry and Physics," *Isis* 66 (March 1975): 75.

149 *on February 23, 1826* Alexander Vucinich, "Nikolai Ivanovich Lobachevskii: The Man behind the First Non-Euclidean Geometry," *Isis* 53 (December 1962): 466.

149 *Lobachevsky became interested in the fifth postulate* P. H. Daus, "The Founding of Non-Euclidean Geometry," *Mathematics News Letter* 7 (April—May 1933): 15; also George Bruce Halsted, "Report on Progress in Non-Euclidean Geometry," *Science* 10 (October 20, 1899): 547.

150 *"imaginary" geometry* Mayme L. Logsdon, "Geometries," *American Mathematical Monthly* 45 (November 1938): 575; also George Bruce Halsted, "The Message of Non-Euclidean Geometry," *Science* 19 (March 11, 1904): 408.

151 *"Nature itself points out distances to us"* Daniels, "Lobachevsky: Some Anticipations of Later Views on the Relation between Geometry and Physics," 78.

151 *Earth, the sun, and Sirius* Ibid., p. 78; also Philip Chapin Jones, "Kant, Euclid, and the Non-Euclideans," *Philosophy of Science* 13 (April 1946): 138.

151 *Mikhail Vasilievich Ostrogradskii* Vucinich, "Nikolai Ivanovich Lobachevskii: The Man behind the First Non-Euclidean Geometry," 467.

154 *the* Tentamen András Prékopa and Emil Molnár, *Non-Euclidean Geometries* (New York: Springer, 2006), 13.

154 *He didn't make any money off the treatise* George Bruce Halsted, "Biography: Bolyai Farkas. [Wolfgang Bolyai]," *American Mathematical Monthly* 3 (January 1896): 2.

155 *1831 was a particularly chaotic time for Gauss* G. Waldo Dunnington, *Gauss: Titan of Science* (Washington DC: Mathematical Association of America, 2004), 137, 143; also Roberto Torretti, *Philosophy of Geometry from Riemann to Poincaré* (Dordrecht, Holland: D. Reidel Publishing Co., 1984), 53.

155 *"Basically I am little touched directly by the local events"* Dunnington, *Gauss: Titan of Science*, 144.

156 *Gauss began to formulate his ideas* Roberto Bonolo, *Non-Euclidean Geometry* (New York: Dover, 1955), 67.

13 "To Praise It Would Be to Praise Myself"

The chapter epigraph is taken from Jeremy Gray, Non-Euclidean Geometry, and the Nature of Space (Cambridge, MA: MIT Press, 2004), p. 54.

157 Eugene Gauss G. Waldo Dunnington, *Gauss: Titan of Science* (Washington, DC: Mathematical Association of America, 2004), 114; also Florian Cajori, "Carl Friedrich Gauss and His Children," *Science* 9 (May 19, 1899): 698.

159 William Gauss Dunnington, *Gauss: Titan of Science*, 204.

159 German immigration Linda Schelbtzki Pickle, *Contented Among Strangers* (Urbana and Chicago: University of Illinois Press, 1996), 11.

160 *a good-for-nothing son* W. K. Bühler, *Gauss: A Biographical Study* (Berlin: Springer-Verlag, 1981), 119.

160 *some eyeglasses* Dunnington, *Gauss: Titan of Science*, 103.

161 *He would relate stories of his childhood* Bühler, *Gauss: A Biographical Study*, 120.

161 *He wrote to his friend Heinrich Schumacher* Roberto Bonolo, *Non-Euclidean Geometry* (New York: Dover, 1955), 67.

161 *"My son is already First Lieutenant"* George Bruce Halsted, "The Foundations of Geometry," *Science* 6 (September 24, 1897): 489.

162 The Science of Absolute Space The full title is taken from Eric W. Weisstein, "Non-Euclidean Geometry," MathWorld—a Wolfram Web Resource,

http://mathworld.wolfram.com/Non-EuclideanGeometry.html); the English translation appears in András Prékopa and Emil Molnár, *Non-Euclidean Geometries* (New York: Springer, 2006), 30.

162   *János had a sense of how important the work was*   Prékopa and Molnár, *Non-Euclidean Geometries*, 4.

162   *The first to recognize the genius of the work*   Roberto Torretti, *Philosophy of Geometry from Riemann to Poincaré*, (Dordrecht, Holland: D. Reidel Publishing Co., 1984), 54; also Jeremy Gray, *Janos Bolyai, Non-Euclidean Geometry, and the Nature of Space* (Cambridge: MIT Press, 2004), 54.

164   *a recurrent fever that may have been caused by malaria*   Prékopa and Molnár, *Non-Euclidean Geometries*, 17; also T. Acs, "Janos Bolyai's Health According to His Military Files," *Orvostort Kolz* 47 (2002): 173–191.

165   *"Here nobody needs mathematics"*   Prékopa and Molnár, *Non-Euclidean Geometries*, 36.

166   Varvara Alexeyavna Moiseyeva   V. Kagan, *N. Lobachevsky: His Contribution to Science* (Moscow: Foreign Language Publishing House, 1957), 66.

14 The Birth of Electronic Communication

The chapter epigraph is taken from G. Waldo Dunnington, *Gauss: Titan of Science* (Washington, DC: Mathematics Association of America, 2004), 147.

168   *great galvanic circuit*   W. K. Bühler, *Gauss: A Biographical Study* (Berlin: Springer-Verlag, 1981), 29; also Dunnington, *Gauss: Titan of Science*, 153.

169   William Weber   Dunnington, *Gauss: Titan of Science*, 139; also G. Waldo Dunnington, "The Sesquicentennial of the Birth of Gauss," *Scientific Monthly* 24 (May 1927): 410; and Bühler, *Gauss: A Biographical Study*, 121.

171   *he began investigating the Earth's magnetic field*   D. J. Struik, "Outline of a History of Differential Geometry II," *Isis* 20 (November 1933): 161–191; also Dunnington, *Gauss: Titan of Science*, 126.

171   *carefully constructed magnetic observatory*   S. R. C. Malin, "Sesquicentenary of Gauss's First Measurement of the Absolute Value of Magnetic Intensity," *Philosophical Transactions of the Royal Society of London, Series A, Mathematical and Physical Sciences* 306, no. 1492, The Earth's Core: Its Structure, Evolution and Magnetic Field. (August 20, 1982): 5; also G. Waldo Dunnington, "The Historical Significance of Carl Friedrich Gauss in Mathematics and Some Aspects of His Work," *Mathematics News Letter* 8 (May 1934): 177.

172   *a model for anyone engaged in measuring the forces of nature*   Carl Friedrich Gauss, *General Investigations of Curved Surfaces*, ed. Peter Pesic (Mineola, NY: Dover, 2005), iii.

172   Intensitas vis Magneticae   Carl Friedrich Gauss, translated from the German by Susan P. Johnson, July 1995, http://21stcenturysciencetech.com/Translations/gaussMagnetic.pdf, last accessed June 15, 2008.

172   magnetic flux density   Wikipedia, http://en.wikipedia.org/wiki/Gauss_(unit), last accessed January 2008.

172   *In his honor*   The definition of a gauss is adapted from Russ Rowlett's Web page at the University of North Carolina at Chapel Hill, www.unc.edu/~rowlett/units/dictG.html, last accessed January 2008.

172   *"I occupy myself now with the Earth's magnetism"*   Malin, "Sesquicentenary of Gauss's First Measurement of the Absolute Value of Magnetic Intensity," 6.

173   *an atlas of measurements*   Dunnington, *Gauss: Titan of Science*, 158.

173   *"All of which formed a brilliant rain of fire"*   Ibid., p. 148.
173   *the Göttingen Seven*   Ibid., pp. 138, 191; also Bühler, *Gauss: A Biographical Study*, 135.
175   *in 1849, Weber returned to Göttingen*   Dunnington, *Gauss: Titan of Science*, 244.
176   *"The earth rarely sees such absolutely pure, noble creatures"*   Bühler, *Gauss: A Biographical Study*, 145.
176   *He began wasting time on séances*   Dunnington, *Gauss: Titan of Science*, 11.

15 The Imaginary Man from Kazan

The chapter epigraph is taken from Alexander Vucinich, "Nikolai Ivanovich Lobachevskii: The Man behind the First Non-Euclidean Geometry," *Isis* 53 (December 1962): 469.

177   *He knew what he had created*   V. Kagan, *N. Lobachevsky: His Contribution to Science* (Moscow: Foreign Language Publishing House, 1957), 50.
177   *"horocycle" and "horosphere"*   Jeremy Gray, *Janos Bolyai, Non-Euclidean Geometry, and the Nature of Space* (Cambridge: MIT Press, 2004), 80.
177   *"A completely new science was created"*   V. Kagan, *N. Lobachevsky: His Contribution to Science*, 50.
177   *he turned farther west*   Vucinich, "Nikolai Ivanovich Lobachevskii: The Man behind the First Non-Euclidean Geometry," 473.
178   *Lobachevsky was a poor writer*   Gray, *Janos Bolyai, Non-Euclidean Geometry, and the Nature of Space*, 78; also G. Waldo Dunnington, *Gauss: Titan of Science* (Washington, DC: Mathematical Association of America, 2004), 186; and Roberto Torretti, *Philosophy of Geometry from Riemann to Poincaré* (Dordrecht, Holland: D. Reidel Publishing Co., 1984), 187.
178   *he started learning Sanskrit*   G. Waldo Dunnington, "The Sesquicentennial of the Birth of Gauss," *Scientific Monthly* 24 (May 1927): 411.
179   *"masterly skill in the true geometrical spirit"*   Vucinich, "Nikolai Ivanovich Lobachevskii: The Man behind the First Non-Euclidean Geometry," 468.
180   *a Hungarian mathematician named Franz Mentovich*   Gray, *Janos Bolyai, Non-Euclidean Geometry, and the Nature of Space*, 80.
182   *there has been more controversy*   András Prékopa and Emil Molnár, *Non-Euclidean Geometries* (New York: Springer, 2006), 34; also George Bruce Halsted, "Report on Progress in Non-Euclidean Geometry," *Science* 10 (October 20, 1899): 545.
183   *a "Golden Jubilee"*   Florian Cajori, "Carl Friedrich Gauss and His Children," *Science* 9 (May 19, 1899): 697.
183   *This was a time of great change*   Michael Rapport, *Nineteenth-Century Europe* (New York: Palgrave Macmillan, 2005), 3.
183   *more than doubled,*   Ibid, p. 5.
184   *Gauss took his last trip away*   Bühler, *Gauss: A Biographical Study*, 154.
184   *a blind, lonely, broke, abandoned, and forgotten man*   Vucinich, "Nikolai Ivanovich Lobachevskii: The Man behind the First Non-Euclidean Geometry," 481.
185   *He had enjoyed a rich career*   Kagan, *N. Lobachevsky: His Contribution to Science*, 65; also Vucinich, "Nikolai Ivanovich Lobachevskii: The Man behind the First Non-Euclidean Geometry," 480; and Alexander Ziwet, "Review of 'Nicolái Ivánovich Lobachévsky' by A. Vasiliev," *Science* 1 (March 29, 1895): 356.
186   *Lobachevsky was not fully appreciated*   "Lobachevsky's contribution to philosophy," *Nature* (June 8, 1957), p. 1176; also Kagan, *N. Lobachevsky: His Contribution to Science*, p. 29.

186   *In 1893, in commemoration of his centenary*   Ziwet, "Review of 'Nicolái Ivánovich Lobachévsky' by A. Vasiliev," 356.

186   *was rediscovered and printed*   P. H. Daus, "The Founding of Non-Euclidean Geometry," *Mathematics News Letter* 7 (April–May 1933): 15.

187   *Farkas Bolyai died the same year*   Prékopa and Molnár, *Non-Euclidean Geometries*, 13.

187   *He was despondent*   Kagan, *N. Lobachevsky: His Contribution to Science*, 25.

187   Baron József Eötvös   Prékopa and Molnár, *Non-Euclidean Geometries*, 41.

189   *Gauss began to grow more ill*   Dunnington, *Gauss: Titan of Science*, 321; also Cajori, "Carl Friedrich Gauss and His Children," 703.

189   Constantine Staufenau   Bühler, *Gauss: A Biographical Study*, 148.

189   *as shrewd with money*   Dunnington, *Gauss: Titan of Science*, 237.

189   *His brain was regarded as*   Ibid., p. 323; also Dunnington, "The Sesquicentennial of the Birth of Gauss," p. 413.

189   *A recent, much more scientific study*   L. F. Haas, "Karl Friedrich Gauss (1777–1855)," *J. Neurol. Neurosurg. Psychiatry* 63 (August 1997); also "Random Samples," *Science* 287 (Feb. 11, 2000): 963.

190   *It took scholars nearly seventy-five years*   Dunnington, *Gauss: Titan of Science*, p. 217.

16 The Soul of the Universe

The chapter epigraph is drawn from an editorial in *American Mathematical Monthly* 2 (September–October 1895): 294.

193   Bernhard Riemann   J. J. O'Connor and E. F. Robertson, "Georg Friedrich Bernhard Riemann," from the Web site of the School of Mathematics and Statistics at the University of St. Andrews, Scotland, www-history.mcs.st-andrews.ac.uk/Biographies/Riemann.html, last accessed July 2007.

194   *Riemann's non-Euclidean geometry*   Roger Osborne, "Some Historic and Philosophic Aspects of Geometry," *Mathematics Magazine* 24, no. 2. (November–December 1950): 79.

194   *He set his sights on the second postulate*   Eric W. Weisstein, "Euclid's Postulates," MathWorld—A Wolfram Web Resource, http://mathworld.wolfram.com/EuclidsPostulates.html, last accessed July 2007.

194   *the fundamental idea that space itself is infinite*   Roberto Torretti, *Philosophy of Geometry from Riemann to Poincaré* (Dordrecht, Holland: D. Reidel Publishing Co., 1984), 40.

195   *"Space is only a special case"*   Quoted in András Prékopa and Emil Molnár, *Non-Euclidean Geometries* (New York: Springer, 2006), 38.

196   *"developed by every conception of the outer world"*   Quoted in Archibald Henderson, "Is the Universe Finite?" *American Mathematical Monthly* 32 (May 1925): 214.

196   *the three types of geometries grew*   Ian Stewart, "Reflections of the past: a review of 'Felix Klein and Sophus Lie' by I.M. Yaglom," *Nature* 334 (July 28, 1988): 306.

196   *the possibility of the finiteness of space*   D. J. Struik, "Outline of a History of Differential Geometry II," *Isis* 20 (November 1933): 175.

196   *"If we assume independence of bodies from position"*   Quoted in Henderson, "Is the Universe Finite?" 214.

196   *By the end of the twentieth century*   Frederick S. Woods, "Space of Constant Curvature," *Annals of Mathematics*, 2nd ser., vol. 3 (1901–1902): 71–112.

196 *A letter that Gauss sent* V. Kagan, *N. Lobachevsky: His Contribution to Science* (Moscow: Foreign Language Publishing House, 1957), 72.

197 Richard Baltzer Roberto Bonolo, *Non-Euclidean Geometry* (New York: Dover, 1955), 123; also George Bruce Halsted, "Biography: Bolyai Farkas. [Wolfgang Bolyai]," *American Mathematical Monthly* 3 (January 1896): 3.

197 Guillaume-Jules Houël Jeremy Gray, *Janos Bolyai, Non-Euclidean Geometry, and the Nature of Space* (Cambridge, MA: MIT Press, 2004), 81.

197 Hermann von Helmholtz Struik, "Outline of a History of Differential Geometry II," 182; also L. Pearce Williams, "The Artful Scientist," *Nature* 378 (November 23, 1995): 345.

197 Eugenio Beltrami Kagan, *N. Lobachevsky: His Contribution to Science*, 72; also Struik, "Outline of a History of Differential Geometry II," 181; and George Bruce Halsted, "Eugenio Beltrami," *American Mathematical Monthly* 9 (March 1902): 60.

198 pseudosphere Hermann Helmholtz, "The Origin and Meaning of Geometrical Axioms," *Mind* 1 (July 1876): 311; also Abe Shenitzer, "How Hyperbolic Geometry Became Respectable," *American Mathematical Monthly* 101 (May 1994): 464–470; and Osborne, "Some Historic and Philosophic Aspects of Geometry," 78.

198 Richard Dedekind Shenitzer, "How Hyperbolic Geometry Became Respectable," 467.

198 Felix Klein Ibid., p. 469; also James Hirschfeld, "Euclidean and Non-Euclidean: a review of 'Foundation of Euclidean and Non-Euclidean Geometries According to Felix Klein' by L. Rédei," *Nature* 219 (August 10, 1968): 658.

199 *"By the weight of his authority"* Carl Friedrich Gauss, *General Investigations of Curved Surfaces,* ed. Peter Pesic (Mineola, NY: Dover, 2005), vii.

199 hyperballein Marvin Jay Greenberg, *Euclidean and Non-Euclidean Geometries: Development and History*, 3rd ed. (New York: W. H. Freeman and Co., 1993), 2.

199 *"They are conventions"* J. W. A. Young, "Review of *La Science et l'Hypothése* by H. Poincaré," *Science* 20 (December 16, 1904): 834.

200 *Thanks in large part to Poincaré's efforts* J. J. O'Connor and E. F. Robertson, "Jules Henri Poincaré," from the Web site of the School of Mathematics and Statistics at the University of St. Andrews, Scotland, www.history.mcs.st-andrews.ac.uk/Biographies/Poincaré.html, last accessed July 2007; also George Bruce Halsted, "Report on Progress in Non-Euclidean Geometry," *Science* 10 (October 20, 1899): 546.

200 *Non-Euclidean geometry allowed artists* Paul M. Laporte, "Cubism and Science," *Journal of Aesthetics and Art Criticism* 7 (March 1949); also Barry Cipra, "Cross-Disciplinary Artists Know Good Math When They See it," *Science* 257 (August 7, 1992): 748; Barry Cipra, "Duchamp and Poincaré Renew an Old Acquaintance," *Science* 286 (November 26, 1999): 1668–1669; Martin Kemp, "Boccioni's ballistics," *Nature* 391 (February 19, 1998); and Martin Kemp, "Dali's Dimensions," *Nature* 391 (January 1, 1998): 27.

200 the Reverend C. L. Dodgson "100 Years Ago," *Nature* 391 (January 22, 1998): 333.

201 *thought of non-Euclidean geometry* Roberto Torretti, "Nineteenth Century Geometry (2003 revision)," *Stanford Encyclopedia of Philosophy*, http://plato.stanford.edu/entries/geometry19th/, last accessed July 2007,

201 *the first contradiction-free mathematical system* Prékopa and Molnár, *Non-Euclidean Geometries*, 47.

201 *All these things are consequences* G. C. McVittie, "Distance and Relativity," *Science* 127 (March 7, 1958): 501–505.

201   *The problem with living on Earth is that we are biased*   Stanley Arthur Eddington, "The Theory of Relativity and Its Influence on Scientific Thought," *Scientific Monthly* 16 (January 1923): 34; also G. C. McVittie, "Distance and Relativity," 505.

202   *The greatest enemy non-Euclidean geometry ever had*   Roberto Torretti, *Philosophy of Physics* (Cambridge, MA: Cambridge University Press, 1999); also T. G. McGonigle, "Euclidean Space: A Lasting Philosophical Obsession," *British Journal for the Philosophy of Science* 21 (May 1970): 185; Philip Chapin Jones, "Kant, Euclid, and the Non-Euclideans," *Philosophy of Science* 13 (April 1946): 137; and Osborne, "Some Historic and Philosophic Aspects of Geometry," p. 81.

202   *his philosophy anticipated non-Euclidean geometry*   J. R. Lucas, "Euclides ab omni naevo vindicatus" [Euclid freed from all flaws], *British Journal for the Philosophy of Science* 20 (May 1969): 1.

202   *there were still staunch opponents at the turn of the century*   Charles N. Moore, "Mathematics and Science," *Science* 81 (January 11, 1935): p. 29.

202   John Lyle   John Lyle, "Queries and Information," *American Mathematical Monthly* 1 (November 1894): 408–410; also Torretti, *Philosophy of Geometry from Riemann to Poincaré*, 33.

203   Eugen Dühring   G. Waldo Dunnington, *Gauss: Titan of Science* (Washington, DC: Mathematical Association of America, 2004), 274.

203   *In the end, non-Euclidean geometry won*   George Bruce Halsted, "Non-Euclidean Geometry," *American Mathematical Monthly* 7 (May 1900): 131; also Judith V. Grabiner, "The Centrality of Mathematics in the History of Western Thought," *Mathematics Magazine* 61 (October 1988): 227; and Richard Trudeau, *The Non-Euclidean Revolution* (Boston: Birkhäuser, 1987), 37.

203   *preposterous—positively hopeless*   George Bruce Halsted, "The Popularization of Non-Euclidean Geometry," *American Mathematical Monthly* 8 (February 1901): 32.

204   *a major advance in abstract thinking*   "The Mathematical Association," *Nature* 195 (July 21, 1962): 234; also Osborne, "Some Historic and Philosophic Aspects of Geometry," 82; and Shenitzer, "How Hyperbolic Geometry Became Respectable," 464.

204   Mileva Maric   Peter Landsberg, "An Uncertainty Principle for Geometry: A Review of 'The Symbolic Universe: Geometry and Physics 1890–1930' edited by Jeremy Gray," *Nature* 404 (April 13, 2000): 705.

17 The Curvature of Space

The chapter epigraph is taken from Stanley Arthur Eddington, "Gravitation and the Principle of Relativity," originally published in 1916 and reprinted in *Nature* 278 (March 15, 1979): 213.

207   *the full eclipse of the sun*   See *Total Solar Eclipse of 1919,* NASA Web site, http://sunearth.gsfc.nasa.gov/eclipse/SEhistory/SE1919May29T.gif, last accessed January 2008.

208   *found them to be non-Euclidean in relativity*   G. C. McVittie, "Distance and Relativity," *Science* 127 (March 7, 1958): 505.

208   *the full equations of electrodynamics*   Carl Friedrich Gauss, *General Investigations of Curved Surfaces,* ed. Peter Pesic (Mineola, NY: Dover, 2005), p. viii.

208   Hendrik Lorentz   J. S. Ames, "Einstein's Law of Gravitation," *Science* 51 (March 12, 1920): 254.

208   *the invariance in certain physical properties*   Elizabeth Hilton, "Aspects of Relativity," *Nature* 205 (February 20, 1965): 735.

208   *that can act on matter*   Abhay Ashtekar, "Physics from geometry," *Nature Physics* 2 (November 2006): 725; also H. J. Ettinger, "Mathematics and the Hypotheses of Science," *National Mathematics Magazine* 11 (November 1936): 75.

208   *is determined by the matter it contains*   H. P. Robertson, "The Expanding Universe," *Science* 76 (September 9, 1932): 223.

209   *"Space tells matter how to move"*   Quoted in Joe Schwartz, "Journey through Relativity," *Nature* 278 (March 15, 1979): 278.

209   *appear as a gravitational field*   Hilton, "Aspects of Relativity," 735.

209   *World War I had visited horrors*   András Prékopa and Emil Molnár, *Non-Euclidean Geometries* (New York: Springer, 2006), 41.

210   *is based on mathematical solutions of the equations of general relativity*   Peter Coles, "The End of the Old Model Universe," *Nature* 393 (June 25, 1998): 741.

210   *"The importance of C. F. Gauss"*   Carl Friedrich Gauss, *General Investigations of Curved Surfaces*, ed. Peter Pesic (Mineola: Dover, 2005): iii.

Bibliography

Acs, T. "Janos Bolyai's Health According to His Military Files." *Orvostort Kolz* 47(2002): 173–191.

American Mathematical Monthly 2 (September–October 1895): 293–294 (editorial).

Ames, J. S. "Einstein's Law of Gravitation." *Science* 51 (March 12, 1920): 253–261.

Archibald, Raymond C. "Gauss and the Regular Polygon of Seventeen Sides." *American Mathematical Monthly* 27 (July–September 1920): 323–326.

———. "History of Mathematics after the Sixteenth Century." *American Mathematical Monthly* 25 (January 1949): 35–56.

Armitage, A. "Foundations of Ancient Science: Review of 'The Exact Sciences in Antiquity' by O. Neugebauer." *Nature* 169 (May 24, 1952): 854.

Artman, Benno. *Euclid: The Creation of Mathematics.* New York: Springer, 1999.

Ashtekar, Abhay. "Physics from Geometry." *Nature Physics* 2 (November 2006): 725–726.

Beeley, P., and C. J. Scriba. *The Correspondence of John Wallis: Volume I (1641–1659).* Oxford: Oxford University Press, 2003.

———. *The Correspondence of John Wallis: Volume II (1660–1668).* Oxford: Oxford University Press, 2005.

Bell, E. T. *Men of Mathematics.* New York: Simon and Schuster, 1986.

———. *The Development of Mathematics.* New York: Dover, 1992.

Bennett, Charles L. "Cosmology from Start to Finish." *Nature* 440 (April 27, 2006): 1126–1131.

Bernstein, Dorothy L. "The Role of Applications in Pure Mathematics." *American Mathematical Monthly* 86 (April 1979): 245–253.

Binzel, Richard P. "Asteroids Come of Age." *Science* 289 (September 22, 2000): 2065–2066.

Birkhoff, George. "The Origin, Nature and Influence of Relativity." *Scientific Monthly* 18 (March 1924): 225–238.

Bonolo, Roberto. *Non-Euclidean Geometry.* New York: Dover, 1955.

Boyer, Carl B. *History of Analytic Geometry.* Mineola, NY: Dover, 2004.

Brown, Ernest W. "The History of Mathematics." *Scientific Monthly* 12 (May 1921): 385–413.

Buchanan, H. E. "The Development of Elementary Geometry." *Mathematics News Letter* 3 (January 1929): 9–18.

Bühler, W. K. *Gauss: A Biographical Study.* Berlin: Springer-Verlag, 1981.

Burch, G. E. "Carl Friedrich Gauss—A Genius Who Apparently Died of Arteriosclerotic Heart Disease and Congestive Heart Failure." *A.M.A. Archives of Internal Medicine* 101 (April 1958): 824–834.

Calinger, Ronald. "Kant and Newtonian Science: The Pre-Critical Period." *Isis* 70 (September 1979): 348–362.

Cajori, Florian. "Attempts Made During the Eighteenth and Nineteenth Centuries to Reform the Teaching of Geometry." *American Mathematical Monthly* 17 (October 1910): 181–201.

———. "Carl Friedrich Gauss and His Children." *Science* 9 (May 19, 1899): 697–704.

Cameron, Edward A. "Review of 'Carl Friedrich Gauss: Titan of Science' by G. Waldo Dunnington." *Scientific Monthly* 83 (October 1956): 205–206.

Cattell, J. McKeen. "The Academy of Sciences." *Science* 16 (December 19, 1902): 965–974.

Cayley, Arthur. "Obligations of Mathematics to Philosophy, and to Questions of Common Life." *Science* 2 (October 5, 1883): 477–483.

Chern, Shiing-Shen. "From Triangles to Manifolds." *American Mathematical Monthly* 86 (May 1979): 339–349.

———. "What Is Geometry?" *American Mathematical Monthly* 97 (October 1990): 679–686.

Cipra, Barry. "Cross-Disciplinary Artists Know Good Math When They See it." *Science* 257 (August 7, 1992): 748–749.

———. "Duchamp and Poincaré Renew an Old Acquaintance." *Science* 286 (November 26, 1999): 1668–1669.

Coles, Peter. "The End of the Old Model Universe." *Nature* 393 (June 25, 1998): 741–744.

"Conformal Mapping: Joukowski Transformation." NASA Glenn Research Center, www.grc.nasa.gov/WWW/K-12/airplane/map.html.

Coxeter, H. S. M. *Non-Euclidean Geometry.* 6th ed. Washington, DC: Mathematical Association of America, 1998.

Crocker, Richard. "Pythagorean Mathematics and Music." *Journal of Aesthetics and Art Criticism* 22 (Spring 1964): 325–335.

Daniels, Norman. "Lobachevsky: Some Anticipations of Later Views on the Relation between Geometry and Physics." *Isis* 66 (March 1975): 75–85.

———. "Thomas Reid's Discovery of a Non-Euclidean." *Philosophy of Science* 39 (June 1972): 219–234.

Daus, P. H. "The Founding of Non-Euclidean Geometry." *Mathematics News Letter* 7 (April–May 1933): 12–16.

"DAWN: A Journey to the Beginning of the Solar System." NASA Fact Sheet, http://dawn.jpl.nasa.gov/mission/dawn_fact_sheet.pdf.

Dehn, Max. "Mathematics, 200 B.C.–600 A.D." *American Mathematical Monthly* 51 (March 1944): 149–157.

De Morgan, Augustus. *A Budget of Paradoxes,* vol. 1. 2nd ed. New York: Dover, 1872; e-book at www.gutenberg.org/files/23100/23100-h/23100-h.htm.

Dicks, D. R. "Thales." *Classical Quarterly* 9 (November 1959): 294–309.

Dolman, John Jr. "Remarks on Professor Lyle's Postulate I. of Euclid's E." *American Mathematical Monthly* 1 (April 1894): 116.

Doolittle, C. L. "Some Advances Made in Astronomical Science during the Nineteenth Century." *Science* 14 (July 5, 1901): 1–12.

Dowling, L. W. "Projective Geometry—Fields of Research." *American Mathematical Monthly* 32 (December 1925): 486–492.

Dunnington, G. Waldo. *Gauss: Titan of Science.* Washington, DC: Mathematical Association of America, 2004.

———. "The Historical Significance of Carl Friedrich Gauss in Mathematics and Some Aspects of His Work." *Mathematics News Letter* 8 (May 1934): 175–179.

———. "The Sesquicentennial of the Birth of Gauss." *Scientific Monthly* 24 (May 1927): 402–414.

Eastman, C. R. "Notes on the History of Natural Science." *Science* 21 (March 31, 1905): 516–517.

Eddington, Arthur Stanley. "Gravitation and the Principle of Relativity." Originally published in 1916 and reprinted in *Nature* 278 (March 15, 1979): 213–214.

———. "The Theory of Relativity and Its Influence on Scientific Thought." *Scientific Monthly* 16 (January 1923): 34–53.

Eells, Walter C. "Discussions: The Ten Most Important Mathematical Books in the World." *American Mathematical Monthly* 30 (September–October 1923): 318–321.

Ettinger, H. J. "Mathematics and the Hypotheses of Science." *National Mathematics Magazine* 11 (November 1936): 71–77.

Euclid, *Elements.* Santa Fe: Green Lion Press, 2003.

———. *Elements—the Bones.* Santa Fe: Green Lion Press, 2002.

Finkel, B. F. "Biography: Karl Frederich Gauss." *American Mathematical Monthly* 8 (February 1901): 25–31.

Franklin, Philip. "The Meaning of Rotation in the Special Theory of Relativity." *Proceedings of the National Academy of Sciences* 8 (September 15, 1922).

Galda, Klaus. "An Informal History of Formal Proofs: From Vigor to Rigor?" *Two-Year College Mathematics Journal* 12 (March 1981): 126–140.

Gauss, Carl Friedrich. *General Investigations of Curved Surfaces.* Edited by Peter Pesic. Mineola, NY: Dover, 2005.

———. *Theory of the Motion of the Heavenly Bodies Moving About the Sun in Conic Sections.* Translated by Charles Henry Davis. Mineola, NY: Dover, 2004.

Grabiner, Judith V. "The Centrality of Mathematics in the History of Western Thought." *Mathematics Magazine* 61 (October 1988): 220–230.

Gray, Jeremy. *Janos Bolyai, Non-Euclidean Geometry, and the Nature of Space.* Cambridge: MIT Press, 2004.

Greenberg, Marvin Jay. *Euclidean and Non-Euclidean Geometries: Development and History.* 3rd ed. New York: W. H. Freeman and Co., 1993.

Guthrie, Kenneth Sylvan. *The Pythagorean Sourcebook and Library.* Grand Rapids, MI: Phanes Press, 1988.

Haas, L. F. "Karl Friedrich Gauss (1777–1855)." *J. Neurol. Neurosurg. Psychiatry* 63 (August 1997).

Halsted, George Bruce. "Bibliography of Hyper-Space and Non-Euclidean Geometry." *American Journal of Mathematics* 1(1878): 261–276.

———. "Biography: Felix Klein." *American Mathematical Monthly* 1 (December 1894): 416–420.

———. "Eugenio Beltrami." *American Mathematical Monthly* 9 (March 1902): 59–63.

———. "Biography: Bolyai Farkas. [Wolfgang Bolyai]." *American Mathematical Monthly* 3 (January 1896): 1–5.

———. "A Class-Book of Non-Euclidean Geometry." *American Mathematical Monthly* 8 (April 1901): 84–87.

———. "The Foundations of Geometry." *Science* 6 (September 24, 1897): 487–491.

———. "Four-Fold Space and Two-Fold Time." *Science* 19 (June 3, 1892): 319.

———. "Gauss and the Non-Euclidean Geometry." *American Mathematical Monthly* 7 (November 1900): 247–252.

———. "A Non-Euclidean Gem." *American Mathematical Monthly* 9 (June–July 1902): 153–159.

———. "Non-Euclidean Geometry." *American Mathematical Monthly* 7 (May 1900): 123–133.

———. "Non-Euclidean Geometry: Historical and Expository." *American Mathematical Monthly* 1 (August 1894): 259–260.

———. "The Message of Non-Euclidean Geometry." *Science* 19 (March 11, 1904): 401–413.

———. "The Popularization of Non-Euclidean Geometry." *American Mathematical Monthly* 8 (February 1901): 31–35.

———. "Report on Progress in Non-Euclidean Geometry." *Science* 10 (October 20, 1899): 545–557.

———. "Simon's Claim for Gauss in Non-Euclidean Geometry." *American Mathematical Monthly* 11 (April 1904): 85–86.

Halsted, George Bruce, et al. "Supplementary Report on Non-Euclidean Geometry." *Science* 14 (November 8, 1901): 705–717.

Harris, D. Fraser. "The Influence of Greece on Science and Medicine." *Scientific Monthly* 3 (July 1916): 51–65.

Haskins, Charles H. "Adelard of Bath." *English Historical Review* 26 (July 1911): 491–498.

———. "Arabic Science in Western Europe." *Isis* 7(1925): 478–485.

Hathaway, A. S. "Non-Euclidean Geometry, or the Science of Absolute Space." *Science* 5 (February 19, 1897): 311–312.

Helmholtz, Hermann. "The Origin and Meaning of Geometrical Axioms." *Mind* 1 (July 1876): 301–321.

Henderson, Archibald. "Is the Universe Finite?" *American Mathematical Monthly* 32 (May 1925): 213–223.

Hilton, Elizabeth. "Aspects of Relativity." *Nature* 205 (February 20, 1965): 735–736.

Hirschfeld, James. "Euclidean and Non-Euclidean: A Review of 'Foundation of Euclidean and Non-Euclidean Geometries According to Felix Klein' by L. Rédei." *Nature* 219 (August 10, 1968): 658.

Hogan, Jenny. "Diary of a Planet's Demise." *Nature* 442 (August 31, 2006): 966–967.

Holden, Warren. "An Attempt to Demonstrate the 11th Axiom of Playfair's Euclid." *American Mathematical Monthly* 2 (May 1895): 146–147.

Jammer, Max. "Geometrical Thinking: Review of 'Space through the Ages. The Evolution of Geometrical Ideas from Pythagoras to Hilbert and Einstein' by Cornelius Lanczos." *Science* 170 (December 11, 1970): 1183.

Jones, Philip Chapin. "Kant, Euclid, and the Non-Euclideans." *Philosophy of Science* 13 (April 1946): 137–143.

Kagan, V. N. *Lobachevsky: His Contribution to Science.* Moscow: Foreign Language Publishing House, 1957.

Kahn, Charles H. *Pythagoras and the Pythagoreans: A Brief History.* Indianapolis: Hackett, 2001.

Kemp, Martin. "Boccioni's Ballistics." *Nature* 391 (February 19, 1998): 751.

———. "Dali's Dimensions." *Nature* 391 (January 1, 1998): 27.

Keyser, C. J. "A Glance at Some Fundamental Aspects of Mathematics." *Scientific Monthly* 6 (June 1918): 481–495.

Kiss, Elemér. *Mathematical Gems from the Bolyai Chest: János Bolyai's Discoveries in Number Theory and Algebra as Recently Deciphered from his Manuscripts.* Budapest: Akadémiai Kiadó, 1999.

Kitcher, Philip. "Kant and the Foundations of Mathematics." *Philosophical Review* 84 (January 1975): 23–50.

Kohn, George C. *The Wordsworth Encyclopedia of Plague and Pestilence.* Hertfordshire, UK: Wordsworth Reference, 1995.

Lanczos, Cornelius. *Space through the Ages: The Evolution of Geometrical Ideas from Pythagoras to Hilbert and Einstein.* New York: Academic Press, 1970.

Landsberg, Peter. "An Uncertainty Principle for Geometry: A Review of 'The Symbolic Universe: Geometry and Physics 1890–1930' edited by Jeremy Gray." *Nature* 404 (April 13, 2000): 705.

Langer, R. E. "Alexandria—Shrine of Mathematics." *American Mathematical Monthly* 48 (February 1941): 109–125.

Laporte, Paul M. "Cubism and Science." *Journal of Aesthetics and Art Criticism* 7 (March 1949): 243–256.

Larmor, Joseph. "Karl Friedrich Gauss and His Family Relatives." *Science* 93 (May 30, 1941): 523–524.

Levine, David B. "Windows into Mathematical Minds: Review of 'Fermat's Enigma: The Epic Quest to Solve the World's Greatest Mathematical Problem' by Simon Singh." *Science* 279 (March 6, 1998): 1485.

Lewis, Florence. "History of the Parallel Postulate." *American Mathematical Monthly* 27 (January 1920): 16–23.

Lissauer, Jack J. "Extrasolar Planets." *Nature* 419 (September 26, 2002): 355–358.

"Lobachevsky's Contribution to Philosophy." *Nature* (June 8, 1957): 1176.

Logsdon, Mayme I. "Geometries." *American Mathematical Monthly* 45 (November 1938): 573–583.

Lucas, J. R. "Euclides ab omni naevo vindicatus." *British Journal for the Philosophy of Science* 20 (May 1969): 1–11.

Lyle, John. "The Angle-Sum According to Playfair." *American Mathematical Monthly* 3 (March 1896): 77–79.

———. "Postulate I. of Euclid's Elements." *American Mathematical Monthly* 1 (January 1894): 11–12.

———. "Queries and Information." *American Mathematical Monthly* 1 (November 1894): 408–410.

———. "True Propositions Not Invalidated by Defective Proofs." *American Mathematical Monthly* 2 (April 1895): 111–112.

Malin, S. R. C."Sesquicentenary of Gauss's First Measurement of the Absolute Value of Magnetic Intensity." *Philosophical Transactions of the Royal Society of London. Series A, Mathematical and Physical Sciences* 306, no. 1492, The Earth's Core: Its Structure, Evolution and Magnetic Field (August 20, 1982): 5–8.

Manning, Henry Parker. *Introductory Non-Euclidean Geometry.* Mineola, NY: Dover, 2005.

Martinez, Jesse A. Fernandez. "Sophie Germain." *Scientific Monthly* 63 (October 1946): 257–260.

"The Mathematical Association." *Nature* (July 21, 1962): 234.

Matz, F. P. "Biography: John Newton Lyle." *American Mathematical Monthly* 3 (April 1896): 95–100.

McGonigle, T. G. "Euclidean Space: A Lasting Philosophical Obsession." *British Journal for the Philosophy of Science* 21 (May 1970): 185–191.

McVittie, G. C. "Distance and Relativity." *Science* 127 (March 7, 1958): 501–505.

Meade, George P. "Youthful Achievements of Great Scientists." *Scientific Monthly* 21 (November 1925): 522–532.

Merton, Robert K. "Priorities in Scientific Discovery: A Chapter in the Sociology of Science." *American Sociological Review* 22 (December 1957): 635–659.

Miller, G. A. "General or Special in the Development of Mathematics." *Science* 93 (March 7, 1941): 235–236.

———. "Some Fundamental Discoveries in Mathematics." *Science* 17 (March 27, 1903): 496–499.

———. "Twenty-Five Important Topics in the History of Secondary Mathematics." *Science* 48 (August 23, 1918): 182–184.

Montgomery, Scott L. "Gained in the translation—Scientific knowledge is enriched as it moves between languages." *Nature* 409 (February 8, 2001): 667.

Moore, Charles N. "Mathematics and Science." *Science* 81 (January 11, 1935): 27–32.

Morehouse, D. W. "Astronomy's Contribution to the Stream of Human Thought." *Science* 75 (January 8, 1932): 27–32.

Murdoch, John E. "Euclides Graeco-Latinus: A Hitherto Unknown Medieval Latin Translation of the Elements Made Directly from the Greek." *Harvard Studies in Classical Philology* 71(1967): 249–302.

Noronha, M. Helena. *Euclidean and Non-Euclidean Geometries.* Upper Saddle River, NJ: Prentice Hall, 2002.

"Obituary Notices of Deceased Fellows." *Proceedings of the Royal Society of London* 7(1954–55): 577–615.

O'Connor, J. J., and E. F. Robertson."Abu Zayd Hunayn ibn Ishaq al-Ibadi." From the Web site of the School of Mathematics and Statistics at the University of St. Andrews, Scotland, www.history.mcs.st-andrews.ac.uk/Biographies/Hunayn .html.

———. "Adelard of Bath." From the Web site of the School of Mathematics and Statistics at the University of St. Andrews, Scotland, www.history.mcs.st-andrews.ac.uk/ Biographies/Adelard.html.

———. "al-Ishbili Abu Muhammad Jabir ibn Aflah." From the Web site of the School of Mathematics and Statistics at the University of St. Andrews, Scotland, www.history .mcs.st-andrews.ac.uk/Biographies/Jabir_ibn_Aflah.html.

———. "Al-Sabi Thabit ibn Qurra al-Harrani." From the Web site of the School of Mathematics and Statistics at the University of St. Andrews, Scotland, www.history .mcs.st-andrews.ac.uk/Biographies/Thabit.html.

———. "Campanus of Novara." From the Web site of the School of Mathematics and Statistics at the University of St. Andrews, Scotland, www.history.mcs.st-andrews .ac.uk/Biographies/Campanus.html.

———. "Claudius Ptolemy." From the Web site of the School of Mathematics and Statistics at the University of St. Andrews, Scotland, www.history.mcs.st-andrews.ac.uk/ Biographies/Ptolemy.html.

———. "Euclid of Alexandria." From the Web site of the School of Mathematics and Statistics at the University of St. Andrews, Scotland, www.history.mcs.st-andrews .ac.uk/Biographies/Euclid.html.

———. "Geminus." From the Web site of the School of Mathematics and Statistics at the University of St. Andrews, Scotland, www.history.mcs.st-andrews.ac.uk/ Biographies/Geminus.html.

———. "Georg Friedrich Bernhard Riemann." From the Web site of the School of Mathematics and Statistics at the University of St. Andrews, Scotland, www.history .mcs.st-andrews.ac.uk/Biographies/Riemann.html.

———. "János Bolyai." From the Web site of the School of Mathematics and Statistics at the University of St. Andrews, Scotland, www.history.mcs.st-andrews.ac.uk/ Biographies/Bolyai.html.

———. "Jules Henri Poincaré." From the Web site of the School of Mathematics and Statistics at the University of St. Andrews, Scotland, www.history.mcs.st-andrews .ac.uk/Biographies/Poincare.html.

———. "Levi ben Gerson." From the Web site of the School of Mathematics and Statistics at the University of St. Andrews, Scotland, www.history.mcs.st-andrews.ac.uk/ Biographies/Levi.html.

———. "Mathematics in Egyptian Papyri." From the Web site of the School of Mathematics and Statistics at the University of St. Andrews, Scotland, www.groups.dcs.st-and.ac.uk/~history/HistTopics/Egyptian_papyri.html.

———. "Nikolai Ivanovich Lobachevsky." From the Web site of the School of Mathematics and Statistics at the University of St. Andrews, Scotland, www.history.mcs .st-andrews.ac.uk/Biographies/Lobachevsky.html.

———. "An overview of Babylonian mathematics." From the Web site of the School of Mathematics and Statistics at the University of St. Andrews, Scotland, www.groups.dcs.st-and.ac.uk/~history/HistTopics/Babylonian_mathematics.html.

———. "An overview of Egyptian mathematics." From the web site of the School of Mathematics and Statistics at University of St. Andrews, Scotland, www.groups .dcs.st-and.ac.uk/~history/HistTopics/Egyptian_mathematics.html.

———. "Robert Recorde." From the Web site of the School of Mathematics and Statistics at the University of St. Andrews, Scotland, www.history.mcs.st-andrews.ac.uk/ Biographies/Recorde.html.

———. "Theon of Alexandria." From the Web site of the School of Mathematics and Statistics at the University of St. Andrews, Scotland, www.history.mcs.st-andrews .ac.uk/Biographies/Theon.html.

"100 Years Ago." *Nature* 391 (January 22, 1998): 333.

Osborne, Roger. "Some Historic and Philosophic Aspects of Geometry." *Mathematics Magazine* 24, no. 2. (November–December 1950): 77–82.

Peirce, C. S. "Campanus." *Science* 13 (May 24, 1901): 809–811.

Peplow, Mark. "Possible Planets Left with No Name." *Nature* 439 (February 9, 2006): 639.

Pickle, Linda Schelbitzki. *Contented Among Strangers.* Urbana and Chicago: University of Illinois Press, 1996.

Pierpont, James. "Geometric Aspects of Einstein's Theory." *Annals of Mathematics*, 2nd ser., vol. 23 (March 1922): 228–254.

———. "The Geometry of Riemann and Einstein." *American Mathematical Monthly* 30 (December 1923): 425–438.

Plackett, R. L. "Studies in the History of Probability and Statistics. XXIX: The Discovery of the Method of Least Squares." *Biometrika* 59 (August 1972): 239–251.

Poincaré, Henri. "Principles of Mathematical Physics." *Scientific Monthly* 82 (April 1956): 165–175.

Prékopa, András, and Emil Molnár. *Non-Euclidean Geometries.* New York: Springer, 2006.

"Random Samples." *Science* 287 (February 11, 2000): 963.

Rapport, Michael. *Nineteenth-Century Europe* (New York: Palgrave Macmillan, 2005).

Reichhardt, Tony. "Space Scientists Get Double Reprieve." *Nature* 440 (March 30, 2006).

Reid, Constance. *A Long Way from Euclid.* Mineola, NY: Dover, 2004.

Richards, Joan L. "The Evolution of Empiricism: Hermann von Helmholtz and the Foundations of Geometry." *British Journal for the Philosophy of Science* 28 (September 1977): 235–253.

Rider, Paul R. "Tenth Annual Meeting of the Missouri Section." *American Mathematical Monthly* 34 (January 1927): 1–4.

Robertson, H. P. "The Expanding Universe." *Science* 76 (September 9, 1932): 221–226.

Russell, C. T., et al. "DAWN: A Journey to the Beginning of the Solar System." ACM Conference Paper, www-ssc.igpp.ucla.edu/dawn/pdf/ACMConferencePaper.

Sack, Robert David. "Geography, Geometry, and Explanation." *Annals of the Association of American Geographers* 62 (March 1972): 61–78.

Salant, William. "Science and Society in Ancient Rome." *Scientific Monthly* 47 (December 1938): 525–535.

Safford, Truman Henry. "Astronomy in the First Half of the Nineteenth Century." *Science* 10 (December 29, 1899): 962–963.

Sarton, George. *Introduction to the History of Science*, vol. 1: *From Homer to Omar Khayyam.* Baltimore: Williams and Wilkins, 1927.

Schilling, Govert. "Underworld Character Kicked Out of Planetary Family." *Science* 313 (September 1, 2006): 1214–1215.

Schroeter, John Jerome. "Observations and Measurements of the Planet Vesta." *Philosophical Transactions of the Royal Society of London* 97 (1807): 245–246.

Schwartz, Joe. "Journey through Relativity." *Nature* 278 (March 15, 1979): 278–279.

"Science News." *Science* 77 (June 2, 1933): 8a–10a.

Scott, J. F. *The Mathematical Work of John Wallis.* New York: Chelsea, 1981.

Seeger, Raymond J. "The Exact Sciences in Antiquity by O. Neugebauer." *Science* 117 (March 6, 1953): 257–258.

Shenitzer, Abe. "How Hyperbolic Geometry Became Respectable." *American Mathematical Monthly* 101 (May 1994): 464–470.

Sheppard, Scott. "A Planet More, A Planet Less." *Nature* 439 (February 2, 2006): 541–542.

Sommerville, D. M. Y. *The Elements of Non-Euclidean Geometry.* Mineola, NY: Dover, 2005.

Spaulding, Edward Gleason. "Are There Any Necessary Truths?" *Journal of Philosophy* 26 (June 6, 1929): 309–329.

Stephenson, J. "The Classification of the Sciences according to Nasiruddin Tusi." *Isis* 5 (1923): 329–338.

Stewart, Ian. "Justifying the Means." *Nature* 354 (November 21, 1991): 184–186.

———. "Reflections of the Past: A Review of 'Felix Klein and Sophus Lie' by I. M. Yaglom." *Nature* (July 28, 1988): 306.

"Striving for Rigor in Greek Science." *Nature* 180 (October 26, 1957): 843–844.

Struik, D. J. "Outline of a History of Differential Geometry II." *Isis* 20 (November 1933): 161–191.

Stump, David. "Poincaré's Thesis of the Translatability of Euclidean and Non-Euclidean Geometries." *Noûs* 25 (December 1991): 639–657.

Struve, Otto. "Freedom of Thought in Astronomy." *Scientific Monthly* 40 (March 1935): 250–256.

Teets, Donald A., and Karen Whitehead. "Computation of Planetary Orbits." *College Mathematics Journal* 29 (November 1998): 397–404.

———. "The Discovery of Ceres: How Gauss Became Famous." *Mathematics Magazine* 72 (April 1999): 83–93.

Thomas, P. C., et al. "Differentiation of the Asteroid Ceres as Revealed by Its Shape." *Nature* 437 (September 8, 2005).

Thorndike, Lynn. "The True Roger Bacon, II." *American Historical Review* 21 (April 1916): 468–480.

Torretti, Roberto. "Nineteenth Century Geometry (2003 revision)." *Stanford Encyclopedia of Philosophy,* http://plato.stanford.edu/entries/geometry-19th/.

———. *Philosophy of Geometry from Riemann to Poincaré.* Dordrecht, Holland: D. Reidel Publishing Co., 1984.

———. *Philosophy of Physics.* Cambridge: Cambridge University Press, 1999.

———. *Relativity and Geometry.* New York: Dover, 1996.

Trudeau, Richard. *The Non-Euclidean Revolution* (Boston: Birkhäuser, 1987).

"UCLA-Led Project Will Send Spacecraft to Study the Origins of the Solar System," UCLA news release, www.universityofcalifornia.edu/news/article/3829.

Veljan, Darko. "The 2500-Year-Old Pythagorean Theorem." *Mathematics Magazine* 73 (October 2000): 259–272.

Vucinich, Alexander. "Nikolai Ivanovich Lobachevskii: The Man behind the First Non-Euclidean Geometry." *Isis* 53 (December 1962): 465–481.

Wallis, John. *The Arithmetic of Infinitesimals.* New York: Springer, 2004.

Weaver, Warren. "Scientific Explanation." *Science* 143 (March 20, 1964): 1297–1300.

Weisstein, Eric W. "Euclid's Postulates." MathWorld—A Wolfram Web Resource, http://mathworld.wolfram.com/EuclidsPostulates.html.

———. "Non-Euclidean Geometry." MathWorld—A Wolfram Web Resource, http://mathworld.wolfram.com/Non-EuclideanGeometry.html.

Whipple, Fred L. "The History of the Solar System." *Proceedings of the National Academy of Science* 52 (August 15, 1964): 565–594.

Whittaker, E. T. "Aristotle, Newton, Einstein." *Science* 98 (September 17, 1943): 249–254.

Williams, L. Pearce. "The Artful Scientist." *Nature* 378 (November 23, 1995): 345.

Wilder, R. L. "The Role of Intuition." *Science* 156 (May 5, 1967): 605–610.

Woodlard, Edgar W. "The Calculation of Planetary Motions." *National Mathematics Magazine* 14 (January 1940): 179–189.

Woods, Frederick S. "Space of Constant Curvature." *Annals of Mathematics*, 2nd ser., vol. 3 (1901–1902): 71–112.

Woodward, R. S. "The Century's Progress in Applied Mathematic." *Science* 11 (January 19, 1900): 81–92.

Young, J. W. A. "Review of *La Science et l'Hypothèse* by H. Poincaré." *Science* 20 (December 16, 1904): 833–837.

Ziwet, Alexander. "Euclid as a Text-Book of Geometry." *Science* 4 (November 7, 1984): 442.

———. "Review of 'Nicolái Ivánovich Lobachévsky' by A. Vasiliev." *Science* 1 (March 29, 1895): 356–358.

Index